Autodesk Maya 2019 for Novices
(Learn By Doing)

CADSoft Technologies

Autodesk Maya 2019 for Novices
Learn By Doing

CADSoft Technologies

Copyright ©2019 by CADSoft Technologies, USA. All rights reserved. Printed in the United States of America except as permitted under the United States Copyright Act of 1976. No part of this publication may be reproduced or distributed in any form or by any means, or stored in the database or retrieval system without the prior permission of CADSOft Technologies.

ISBN 978-1-64057-070-2

NOTICE TO THE READER

Publisher does not warrant or guarantee any of the products described in the text or perform any independent analysis in connection with any of the product information contained in the text. Publisher does not assume, and expressly disclaims, any obligation to obtain and include information other than that provided to it by the manufacturer.

The reader is expressly warned to consider and adopt all safety precautions that might be indicated by the activities herein and to avoid all potential hazards. By following the instructions contained herein, the reader willingly assumes all risks in connection with such instructions.

The Publisher makes no representation or warranties of any kind, including but not limited to, the warranties of fitness for particular purpose or merchantability, nor are any such representations implied with respect to the material set forth herein, and the publisher takes no responsibility with respect to such material. The publisher shall not be liable for any special, consequential, or exemplary damages resulting, in whole or part, from the reader's use of, or reliance upon, this material.

DEDICATION

To students, who are dedicated to learning new technologies

THANKS

*To employees of CADSoft Technologies
for their cooperation*

Online Training Program Offered by CADSoft Technologies

CADSoft Technologies provides effective and affordable virtual online training on various software packages including Computer Aided Design, Manufacturing, and Engineering (CAD/CAM/CAE), computer programming languages, animation, architecture, and GIS. The training is delivered 'live' via Internet at any time, any place, and at any pace to individuals as well as the students of colleges, universities, and CAD/CAM/CAE training centers. The main features of this program are:

Training for Students and Companies in a Classroom Setting

A team of highly experienced instructors and qualified engineers at CADSoft Technologies conduct the classes in classroom environment at its centers. This team has authored several textbooks that are rated "one of the best" in their categories and are used in various colleges, universities, and training centers in North America, Europe, and in other parts of the world.

CADSoft Technologies with its cost effective and time saving initiative strives to deliver online training in the comfort of your home or work place, thereby relieving you from the hassles of traveling to training centers. CADSoft Technologies strives to be the best training institute in USA. CADSoft Technologies has developed career oriented programs by following state-of-the-art teaching and learning methodologies.

Training for Individuals

CADSoft Technologies with its cost effective and time saving initiative strives to deliver the training in the comfort of your home or work place, thereby relieving you from the hassles of traveling to training centers.

Training Offered on Software Packages

CADSoft provides basic and advanced training on the following software packages:

CAD/CAM/CAE: CATIA, Pro/ENGINEER Wildfire, Creo Parametric, Creo Direct, SOLIDWORKS, Autodesk Inventor, Solid Edge, NX, AutoCAD, AutoCAD LT, AutoCAD Plant 3D, Customizing AutoCAD, EdgeCAM, and ANSYS

Architecture and GIS: Autodesk Revit (Architecture, Structure, MEP), AutoCAD Civil 3D, AutoCAD Map 3D, Navisworks, Oracle Primavera, and Bentley STAAD Pro

Animation and Styling: Autodesk 3ds Max, Autodesk Maya, Blender, Pixologic ZBrush, and MAXON CINEMA 4D

Computer Programming: C++, VB.NET, Oracle, AJAX, and Java

Personality Development: Personality Development and Engineering Ethics/Soft Skills Course

Table of Contents

Dedication iii
Preface xvii

Chapter 1
Exploring Maya Interface..1-1

Chapter 2
Polygon Modeling..2-1

Chapter 3
NURBS Curves and Surfaces..3-1

Chapter 4
NURBS Modeling...4-1

Chapter 5
UV Mapping..5-1

Chapter 6
Shading and Texturing...6-1

Chapter 7
Lighting...7-1

Chapter 8
Animation...8-1

Chapter 9
Rigging, Constraints, and Deformers..9-1

Chapter 10
Paint Effects..10-1

Chapter 11
Rendering... 11-1

Index I-1

Preface

Autodesk Maya 2019

Welcome to the world of Autodesk Maya 2019. Autodesk Maya 2019 is a powerful, integrated 3D modeling, animation, visual effects, and rendering software developed by Autodesk Inc. This integrated node-based 3D software finds its application in the development of films, games, and design projects. A wide range of 3D visual effects, computer graphics, and character animation tools make it an ideal platform for 3D artists. The intuitive user interface and workflow tools of Maya 2019 have made the job of design visualization specialists a lot easier.

Autodesk Maya 2019 for Novices textbook covers all features of Autodesk Maya 2019 in a simple, lucid, and comprehensive manner. It aims at harnessing the power of Autodesk Maya 2019 for 3D and visual effects artists, and designers. This textbook will help you transform your imagination into reality with ease. Also, it will unleash your creativity, thus helping you create realistic 3D models, animation, and visual effects. It caters to the needs of both the novice and advanced users of Maya 2019 and is ideally suited for learning at your convenience and at your pace.

The salient features of this textbook are as follows:

- **Tutorial Approach**
  The author has adopted the tutorial point-of-view and the learn-by-doing approach throughout the textbook. This approach will guide the users through the process of creating the models, adding textures, and animating them in the tutorials.

- **Real-World Models as Projects**
  The author has used about 37 real-world modeling and animation projects as tutorials in this textbook. This will enable the readers to relate the tutorials to the real-world models in the animation and visual effects industry. In addition, there are about 34 exercises that are also based on the real-world animation projects.

- **Tips and Notes**
  Additional information related to various topics is provided to the users in the form of tips and notes.

- **Learning Objectives**
  The first page of every chapter summarizes the topics that will be covered in that chapter.

- **Exercises**
  Exercises are given at the end of the chapter and they can be used by the instructors to assess the knowledge of the student. You need to perform all the exercises as output of some of these exercises may have been used as a reference later in this book.

Symbols Used in the Textbook

Note

The author has provided additional information to the users about the topic being discussed in the form of notes.

Tip

Special information and techniques are provided in the form of tips that helps in increasing the efficiency of the users.

Formatting Conventions Used in the Textbook

Please refer to the following list for the formatting conventions used in this textbook.

- Names of tools, buttons, options, tabs, attributes, renderer, and toolbars are written in bold face

    Example: The **Unfold** tool, the **Apply and Close** button, the **Assign Material to Selection** option, the Maya Software renderer, the **Fill Style** attribute, and so on.

- Names of dialog boxes, drop-down lists, areas, edit boxes, check boxes, and radio buttons are written in boldface.

    Example: The **Save As** dialog box, the **Look In** drop-down list, the **Display** area, the **Particle name** edit box, the **Color feedback** check box, and the **Center** radio button.

- Values entered in edit boxes are written in boldface.

    Example: In the **Particle Size** area, enter the value **0.450** in the **Radius** edit box.

- Names of the files are italicized.

    Example: *c13tut2.mb*

- The methods of invoking a tool/option from menubar or the toolbar are given in a shaded box.

    Menubar: Edit Mesh > Components > Bevel
    Panel Toolbar: Select camera tool

Naming Conventions Used in the Textbook

Tool
If you click on an item in a panel of the Tool Box and a command is invoked to create/edit an object or perform some action, then that item is termed as **tool**.

For example:
Select Tool, **Lasso Tool**, **Move Tool**, **Scale Tool**, **Rotate Tool**, **Show Manipulator Tool**

Flyout
A flyout is a menu that contains options with similar type of functions. Figure 1 shows the flyout displayed on pressing the right mouse button on the **Select camera** tool.

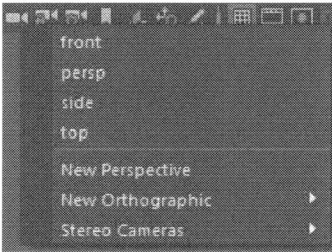

Figure 1 The flyout displayed on clicking the right mouse button on the **Select camera** tool

Marking Menus
Marking menus are similar to shortcut menus that consist of almost all the tools required to perform an operation on an object. There are three types of marking menus in Maya.

The first type of marking menu is used to create default objects in the viewport. To create a default object, press and hold the SHIFT key and then right-click anywhere in the viewport; a marking menu will be displayed, as shown in Figure 2.

The second type of marking menu is used to switch among various components of an object such as vertices, faces, edges, and so on. To invoke this marking menu, select an object and right-click; a marking menu will be displayed, as shown in Figure 3.

The third type of marking menu is used to modify the components of an object. To invoke this marking menu, select a component, press and hold the SHIFT key, and then right-click on the selected object; a marking menu will be displayed, as shown in Figure 4.

Preface

ix

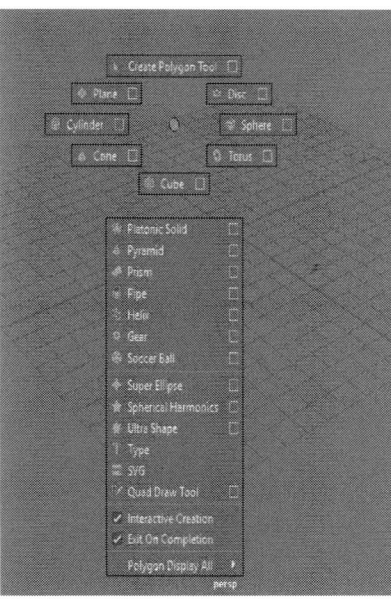

Figure 2 *Marking menu displaying options for creating default objects*

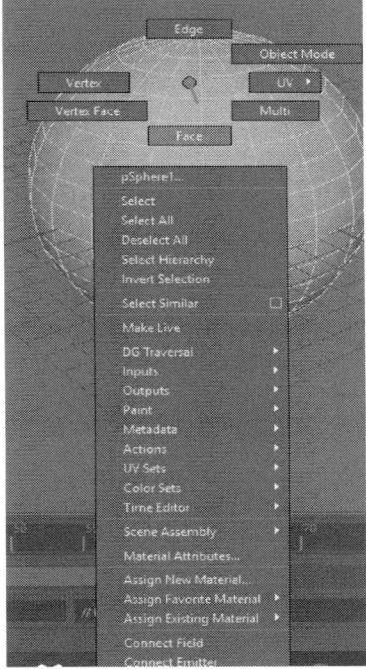

Figure 3 *Marking menu displaying components of the selected object*

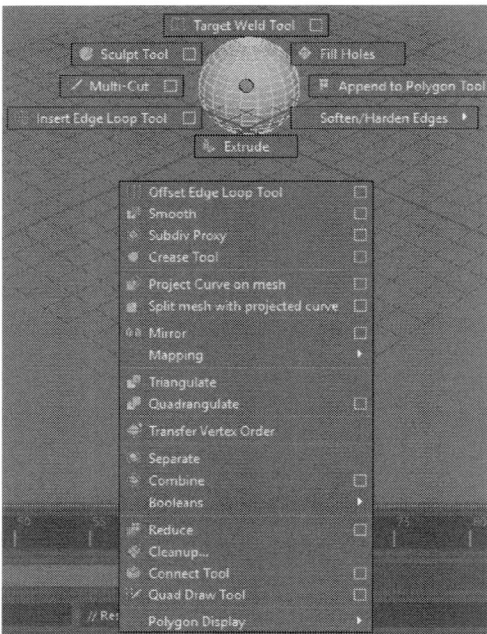

Figure 4 *The marking menu displaying various tools for modifying the components of an object*

Button

The item in a dialog box that has a 3D shape is termed as **Button**. For example, **Extrude** button, **Apply** button, **Close** button, and so on, refer to Figure 5.

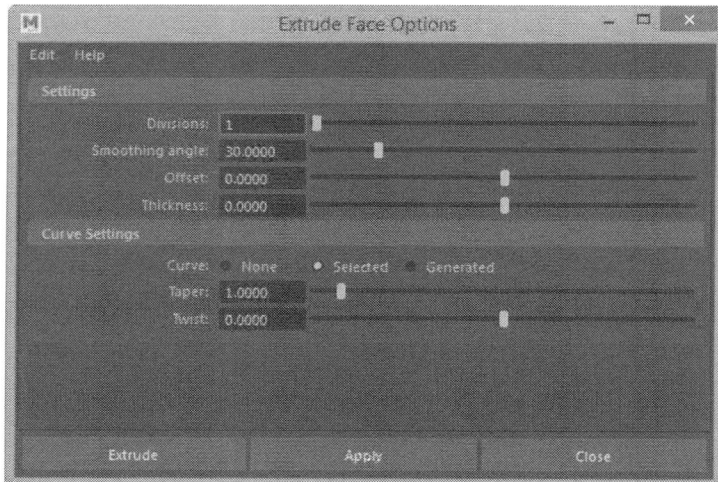

Figure 5 *The **Extrude**, **Apply**, and **Close** buttons*

Drop-down List

A drop-down list is the one in which a set of options are grouped together. You can set various parameters using these options. You can identify a drop-down list with a down arrow on it. For example, **Menuset** drop-down list, refer to Figure 6.

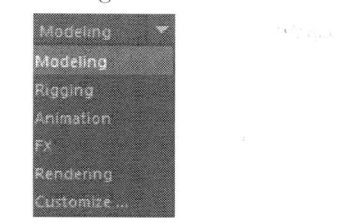

Figure 6 *The **Menuset** drop-down list*

This page is intentionally left blank

Chapter 1

Exploring Maya Interface

Learning Objectives

After completing this chapter, you will be able to:
- *Start Autodesk Maya 2019*
- *Work with menusets in Autodesk Maya*
- *Understand various terms related to Maya interface*

INTRODUCTION TO Autodesk Maya

Welcome to the world of Autodesk Maya 2019. Maya is a 3D software, developed by Autodesk Inc., which enables you to create realistic 3D models and visual effects with much ease. Although Maya is quite a vast software to deal with, yet all the major tools and features used in Autodesk Maya 2019 have been covered in this book.

STARTING Autodesk Maya 2019

To start Autodesk Maya 2019, double-click on the shortcut icon of Autodesk Maya 2019 displayed on the desktop of your computer, as shown in Figure 1-1. This icon is automatically created on installing Autodesk Maya 2019 on your computer.

Figure 1-1 Starting Autodesk Maya 2019 by choosing the icon from desktop

Double-click on the icon; three windows namely, **Output Window**, the main **Autodesk Maya 2019** interface window, and the **What's New Highlight Settings** window will be displayed on the screen. The **Output Window** is shown in Figure 1-2. By default, all the new tools and icons are highlighted in green in Maya 2019. The **What's New Highlight Settings** window, as shown in Figure 1-3, is used to toggle the visibility of these highlights.

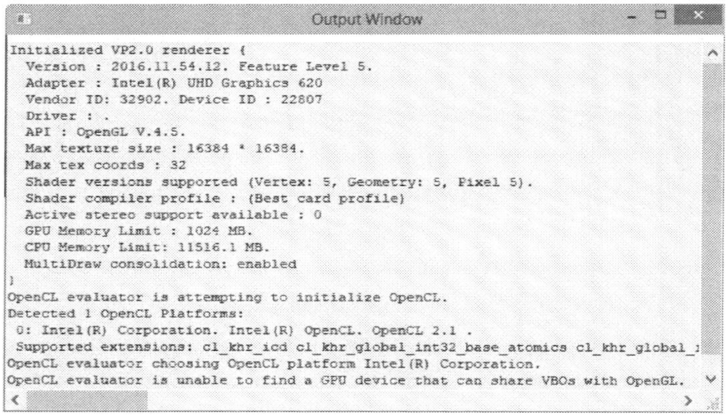

Figure 1-2 The **Output Window**

Exploring Maya Interface 1-3

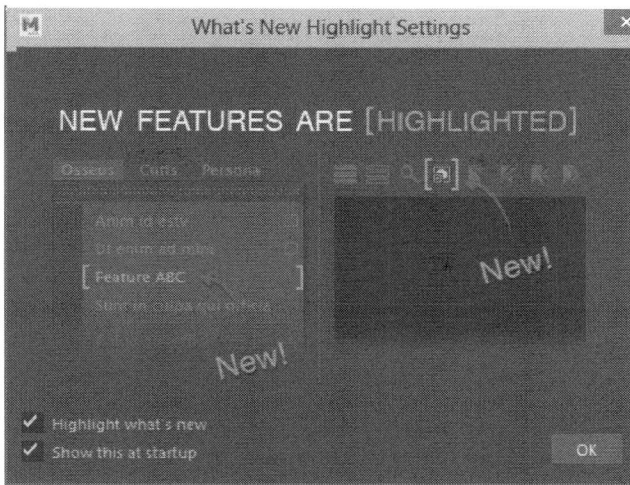

Figure 1-3 The **What's New Highlight Settings** window

AUTODESK MAYA 2019 SCREEN COMPONENTS

Autodesk Maya interface consists of viewports, title bar, menubar, Status Line, Shelf, Tool Box, and so on. All these components will be discussed later in this chapter. When you start Autodesk Maya 2019 for the first time, the persp viewport is displayed by default, refer to Figure 1-4. Workspace is the part or the work area where you can create a 3D scene. Workspaces are also known as viewports or views. In this textbook, the workspaces will be referred to as viewports. Every viewport has a grid placed in the center. The grid acts as a reference that is used in aligning the 3D objects or 2D curves. A grid is a pattern of straight lines that intersect with each other to form squares. The center of the grid is intersected by two dark lines. The point of intersection of these two dark lines is known as the origin. The origin is an arbitrary point, which is used to determine the location of the objects. All the three coordinates, X, Y, and Z are set at 0 position on the origin. Note that in Maya, the X, Y, and Z axes are displayed in red, green, and blue colors, respectively.

Figure 1-4 The default interface of Autodesk Maya 2019 with persp viewport displayed

Autodesk Maya 2019 is divided into four viewports: top-Y, front-Z, side-X, and persp. These viewports are classified into two categories, orthographic, and isometric. The orthographic category comprises the top, front, and side viewports and the isometric category consists of the persp viewport. The orthographic viewport displays the 2-dimensional (2D) view of the objects created in it, whereas the isometric viewport displays the 3-dimensional (3D) view of the objects created. Every viewport can be recognized easily by its name, which is displayed at the bottom of each viewport. Figure 1-5 shows various components of the Maya interface.

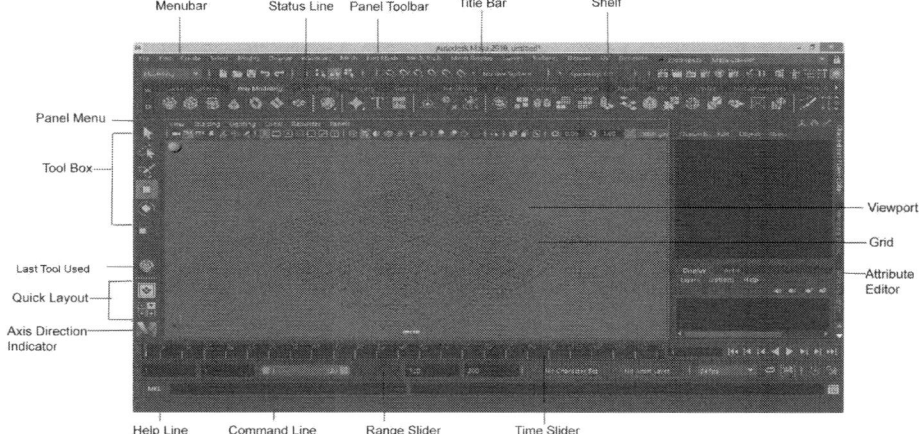

Figure 1-5 *Displaying various screen components of the Maya interface*

Every viewport has its own **Panel** menu that allows you to access the tools related to that specific viewport. The Axis Direction Indicator located at the lower left corner of each viewport indicates about the X, Y, and Z axes. Similarly, every viewport in Maya has a default camera applied to it through which the viewport scene is visible. The name of the camera is displayed at the bottom of each viewport. In other words, the name of the viewport is actually the name of the camera of that particular viewport.

The title bar, which lies at the top of the screen, displays the name and version of the software, the name of the file, and the location where the file is saved. A Maya file is saved with the *.mb* or *.ma* extension. The three buttons on the extreme right of title bar are used to minimize, maximize, and close the Autodesk Maya 2019 window, respectively. Various interface components of the Autodesk Maya 2019 interface are discussed next.

Tip
To toggle between single viewport and four viewport views, hover the cursor over one of the viewports and press the SPACEBAR key.

Exploring Maya Interface

Menubar

The menubar is available just below the title bar. The type of menubar displayed depends on menusets. In Maya, there are different menusets namely, **Modeling**, **Rigging**, **Animation**, **FX**, and **Rendering**. These menusets are displayed in the **Menuset** drop-down list located on the extreme left of the Status Line. On selecting a particular menuset, the menus in the menubar change accordingly. However, there are nine common menus in Maya that remain constant irrespective of the menuset chosen. Figure 1-6 shows the menubar corresponding to the **Modeling** menuset.

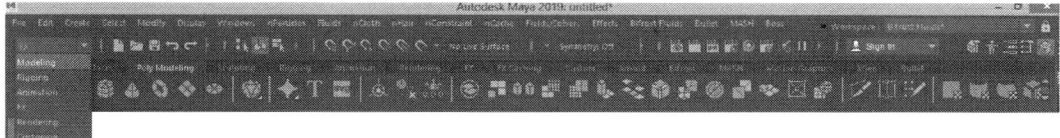

*Figure 1-6 Menubar displayed on choosing the **Modeling** menuset*

On invoking a menu from the menubar, a pull-down menu is displayed. On the right of some of the options in these pull-down menus, there are two types of demarcations, arrows and option boxes. When you click on an option box, a window will be displayed. You can use this window to set the options for that particular tool or menu item. On clicking the arrow, the corresponding cascading menu will be displayed.

Tip
You can also select different menusets using the hotkeys that are assigned to them. The default hotkeys are F2 (Modeling), F3 (Rigging), F4 (Animation), F5 (FX), and F6 (Rendering).

Status Line

The Status Line is located below the menubar. The **Menuset** drop-down list is located at the left of the Status Line. The Status Line consists of different graphical icons. The graphical icons are further grouped and these groups are separated by vertical lines with either a box or an arrow symbol in the middle. These vertical lines are known as Show/Hide buttons, refer to Figure 1-7. You can click on a Show/Hide button with a box symbol to hide particular icons on the Status Line. On doing so, the corresponding icons will hide and the box will change into an arrow symbol. Similarly, if you click on a Show/Hide button that has an arrow symbol in the middle, the icons of the corresponding group will be displayed. Various groups separated by Show/Hide buttons are discussed next.

Figure 1-7 The Status Line

Menuset

As mentioned earlier, the **Menuset** drop-down list in the Status Line has different menusets such as **Modeling**, **Rigging**, **Animation**, **FX**, and **Rendering**, as shown in Figure 1-8. The options displayed in the menubar depend upon the menuset selected from this drop-down list. For example, if you select the **Rendering** menuset from the

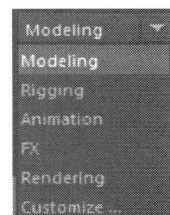

*Figure 1-8 The **Menuset** drop-down list*

Menuset drop-down list, all the commands related to it will be displayed in the menus of the menubar. You can add a custom menuset by selecting the **Customize** option. On selecting it, the **Menu Set Editor** window will be displayed, as shown in Figure 1-9. To create a new menuset, choose the **New Menu Set** button from this window; the **Create New Menu Set** window will be displayed. Enter the menu name in **Enter name** edit box and then choose the **Create** button, the new menuset will be added in the **Menu sets** area of the window. To add a menu in the **Menus in menu set** area; select the desired menu items from the **All menus** area and right-click on it. Next, choose **Add to Menu Set** from the shortcut menu displayed; the selected menu items will be added to the **Menus in menu set** area. Now, choose the **Close Window** button to close the window.

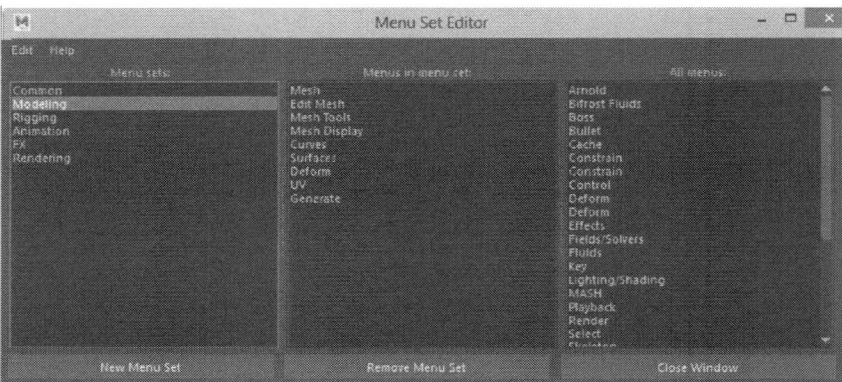

Figure 1-9 The **Menu Set Editor** window

Shelf

The Shelf is located below the Status Line, as shown in Figure 1-33. The Shelf is divided into two parts. The upper part in the Shelf consists of different Shelf tabs and lower part displays the icons of different tools. The icons displayed in this area depend on the tab chosen, refer to Figure 1-10.

Figure 1-10 The Shelf

You can also customize the Shelf as per your requirement. To do so, press and hold the left mouse button over the **Menu of items to modify the shelf** button, refer to Figure 1-11; a flyout will be displayed, as shown in Figure 1-11. Various options in this flyout are discussed next.

Shelf Tabs

The **Shelf Tabs** option is used to toggle the visibility of the Shelf tabs. On choosing this option, the Shelfs tabs will disappear, and only the tool icons corresponding to the selected tab will be visible.

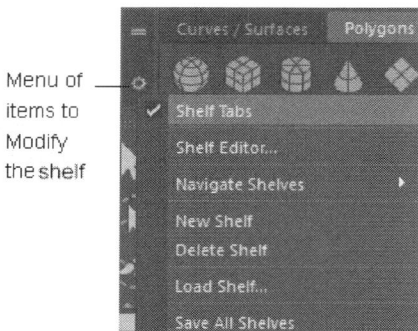

Figure 1-11 Flyout displayed on choosing the **Menu of items to modify the shelf** button

Exploring Maya Interface

Shelf Editor

The **Shelf Editor** option is used to create a Shelf and edit the properties of an existing Shelf. When this option is chosen, the **Shelf Editor** will be displayed in the viewport, as shown in Figure 1-12. Alternatively, you can choose **Windows > Editors > Settings/Preferences > Shelf Editor** from the menubar to display the **Shelf Editor**. In the **Shelf Editor**, you can change the name and position of shelves and their contents. You can also create a new shelf and its contents using the **Shelf Editor**.

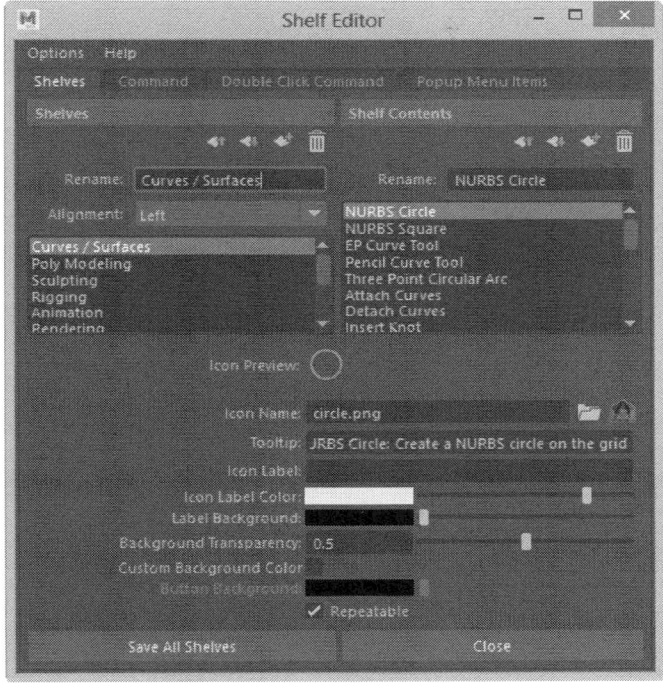

Figure 1-12 The **Shelf Editor**

Navigate Shelves

The **Navigate Shelves** option is used to choose the previous or next Shelf of the currently chosen Shelf. On choosing this option, a cascading menu will be displayed, as shown in Figure 1-13. The options in the cascading menu are discussed next.

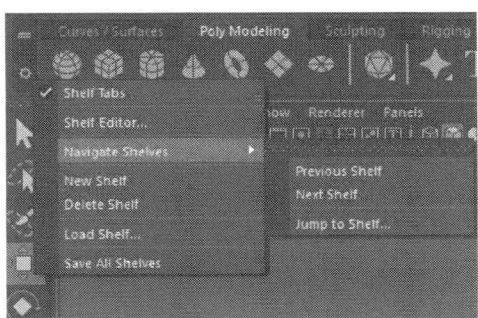

Figure 1-13 Cascading menu displayed on choosing the **Navigate Shelves** option

Tool Box

The Tool Box is located on the left side of the workspace. It comprises of the most commonly used tools in Maya. In addition to the commonly used tools, the Tool Box has several other options or commands that help you change the layout of the interface. Various tools in the Tool Box are discussed next.

Select Tool

The **Select Tool** is used to select the objects created in the viewport. To select an object, invoke the **Select Tool** from the Tool Box and click on an object in the viewport; the object will be selected. On invoking this tool, the manipulators will not be activated.

Lasso Tool

The **Lasso Tool** is used to select an object by using a free hand marquee selection. This tool is very much similar to the **Select Tool**. To select an object, invoke the **Lasso Tool**; the cursor will change to a rope knot. Next, press and hold the left mouse button and drag the cursor in the viewport to create a selection area around the object. Then, release the left mouse button; the object inside the selection area will be selected. To adjust the properties of the **Lasso Tool**, make sure that the **Lasso Tool** is invoked, and then choose the **Show/Hide the Tool Settings** button from the Status Line; the **Tool Settings (Lasso Tool)** window will be displayed. Adjust the **Lasso Tool** properties from the **Tool Settings (Lasso Tool)** window as per your requirement.

Paint Selection Tool

The **Paint Selection Tool** is used to select various components of an object. To select various components of an object, invoke the **Select Tool** from the Tool Box and select an object in the viewport. Next, press and hold the right mouse button over the selected object; a marking menu will be displayed. Choose **Vertex** from the marking menu to make the vertex selection mode active. Now, choose the **Paint Selection Tool** from the Tool Box; the cursor will change to the paint brush. Next, press and hold the left mouse button and drag the cursor over the object to select the desired vertices. To go back to the object mode, invoke the **Select Tool** and then press and hold the right mouse button; a marking menu will be displayed. Choose **Object Mode** from the marking menu to make the vertex selection mode inactive.

You can also increase the size of the **Paint Selection Tool** cursor. To do so, press and hold the B key on the keyboard. Next, press and hold the left mouse button in the viewport and drag the cursor to adjust the size of the brush.

Move Tool

The **Move Tool** is used to move an object from one place to another in the viewport. To do so, invoke **Move Tool** from the Tool Box; the cursor will change to an arrow with a box at its tip. Select the object in the workspace that you want to move. You can move the selected object in the X, Y, and Z directions by using the handles/manipulators over the object. You can also adjust the properties of the **Move Tool** by choosing the **Show or Hide the Tool Settings** button from the Status Line or by double-clicking on the **Move Tool** itself. To use the **Move Tool**, you need to create an object in the viewport. To do so, create a sphere by choosing **Create > Objects > Polygon Primitives > Sphere** from the menubar.

Exploring Maya Interface

A sphere will be created. Now, invoke the **Move Tool** from the Tool Box and select the object created by clicking on it; the **Move Tool** manipulator will be displayed on the selected object with three color handles, as shown in Figure 1-14. These three color handles are used to move the object in the X, Y, or Z direction. The colors of the handles represent three axes; red represents the X-axis, green represents the Y-axis, and blue represents the Z-axis. At the intersection point of these handles, a box will be displayed that can be used to move the object proportionately in all the three directions. Press and hold the left mouse button over the box and drag the cursor to move the object freely in the viewport. To adjust the default settings of the **Move Tool**, double-click on it in the Tool Box; the **Tool Settings (Move Tool)** window will be displayed, as shown in Figure 1-15. Change the settings as per your requirement in this window.

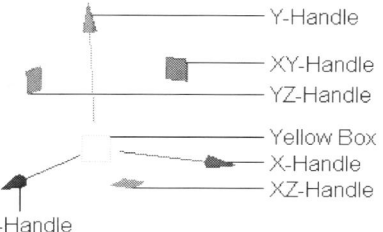

*Figure 1-14 The **Move Tool** manipulator*

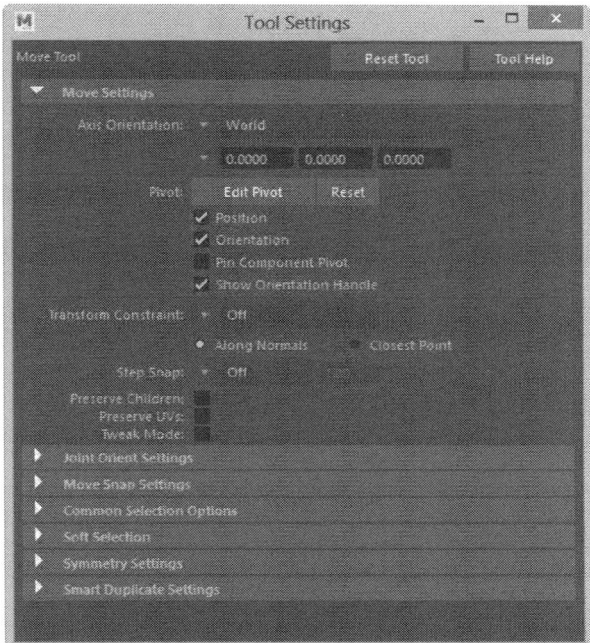

*Figure 1-15 The **Tool Settings (Move Tool)** window*

By default, the pivot point is located at the center of the object. To change the pivot point, make sure that the **Move Tool** is invoked and then press the INSERT key; the pivot point will be displayed in the viewport, as shown in Figure 1-16. Move the pivot point to adjust its position. You can also put the pivot at the center of the object. To do so, choose **Modify > Pivot > Center Pivot** from the menubar; the pivot point will be adjusted to the center of the object. You can also adjust the pivot point by pressing and holding the D key and moving the manipulator.

Note
A pivot is a point in 3D space that is used as a reference point for the transformation of objects.

Rotate Tool

The **Rotate Tool** is used to rotate an object along the X, Y, or Z axis. To rotate an object in the viewport, select the object and invoke the **Rotate Tool** from the Tool Box; the **Rotate Tool** manipulator will be displayed on the object, as shown in Figure 1-17. The **Rotate Tool** manipulator consists of three colored rings. The red ring represents the X axis, whereas the green and blue rings represent the Y and Z axes, respectively. Moreover, the yellow ring around the selected object helps you rotate the selected object in the view axis. On selecting a particular ring, its color changes to yellow. You can change the default settings of the **Rotate Tool** by double-clicking on it in the Tool Box. On doing so, the **Tool Settings (Rotate Tool)** window will be displayed, as shown in Figure 1-18. This window contains various options for rotation. You can change the settings in this window as required.

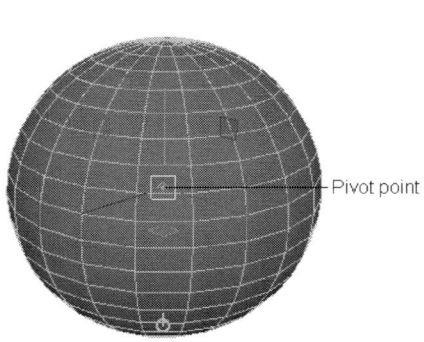

Figure 1-16 The pivot point

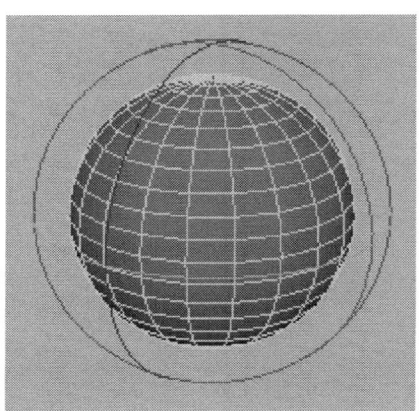

Figure 1-17 The **Rotate Tool** manipulator

Scale Tool

The **Scale Tool** is used to scale an object along the X, Y, or Z-axis. To scale an object in the viewport, select the object and invoke **Scale Tool** from the Tool Box; **Scale Tool** manipulator will be displayed on the object, as shown in Figure 1-19.

The **Scale Tool** manipulator consists of three boxes. The red box represents the X axis, whereas the green and blue boxes represent the Y and Z axes, respectively. Moreover, the yellow colored box in the center lets you scale the selected object uniformly in all axes. On selecting any one of these colored scale boxes, the default color of the box changes to yellow. You can also adjust the default settings of **Scale Tool** by double-clicking on it in the Tool Box. On doing so, the **Tool Settings (Scale Tool)** window will be displayed, as shown in Figure 1-20. Make the required changes in the **Tool Settings (Scale Tool)** window to adjust the basic attributes of **Scale Tool**.

Exploring Maya Interface

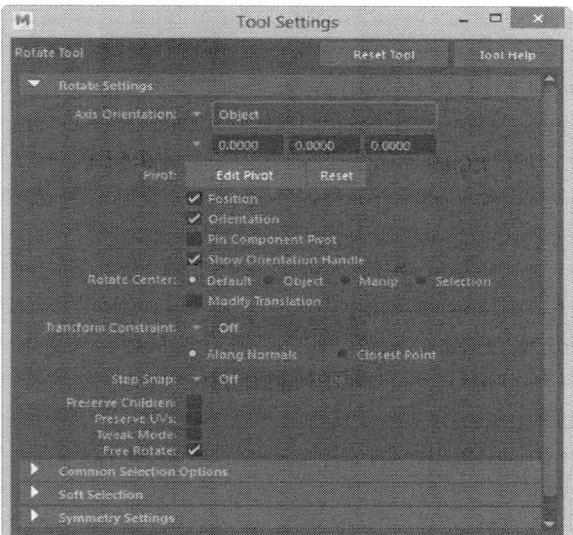

Figure 1-18 Partial view of the **Tool Settings** (**Rotate Tool**) *window*

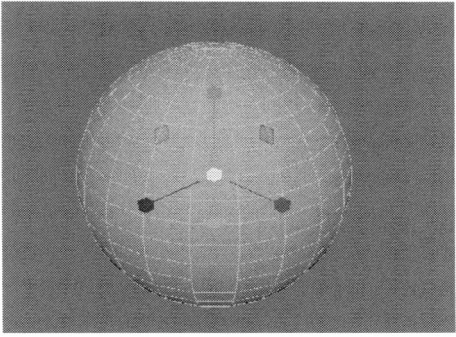

Figure 1-19 The **Scale Tool** *manipulator*

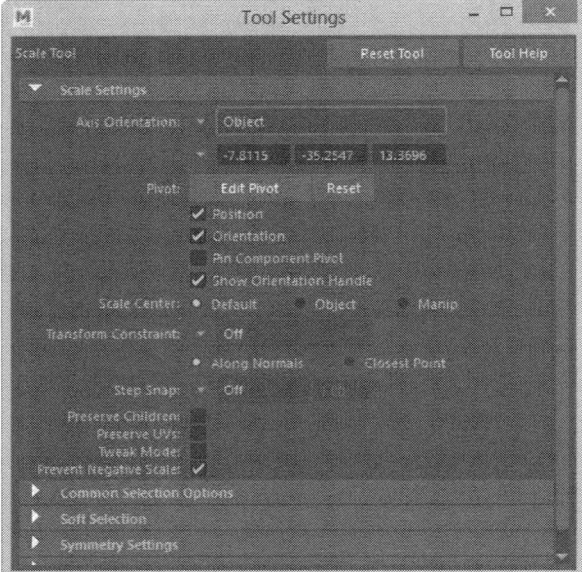

*Figure 1-20 Partial view of the **Tool Settings (Scale Tool)** window*

Note
While rotating, moving, or scaling an object, different colored handles are displayed. These handles indicate different axes. You can use this color scheme while working with three transform tools as well. The red, green, and blue colors represent the X, Y, and Z axes, respectively.

Last Tool Used

The **Last Tool Used** tool is used to invoke the last used or the currently selected tool. This tool displays the icon of the last used tool or currently active tool.

Quick Layout Buttons

Using the buttons in the Quick Layout buttons area, refer to Figure 1-5, you can the toggle the display of layouts as required. You can also change the display of layout buttons. To do so, right-click on one of the Quick Layout buttons; a shortcut menu with various layout options will be displayed, as shown in Figure 1-21. Next, choose any of the layout from the shortcut menu as per your need; the current layout will be replaced by the chosen layout. Using these buttons, you can also edit the current layout. To do so, right-click on the Quick Layout buttons; a shortcut menu will be displayed. Choose **Edit Layouts** from the shortcut menu; the **Panels** window will be displayed, as shown in Figure 1-22.

Exploring Maya Interface

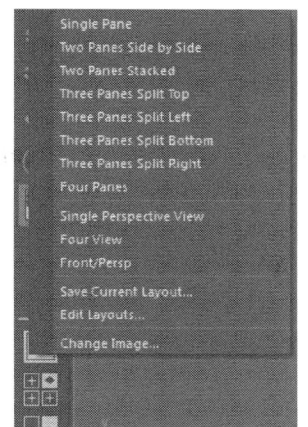

Figure 1-21 *A shortcut menu with various layout*

Time Slider and Range Slider

The Time Slider and the Range Slider, as shown in Figure 1-23, are located at the bottom of the viewport. These two sliders are used to control the frames in animation. The Time Slider comprises of the frames that are used for animation. There is an input box on the Time Slider called **Set the current time**, which indicates the current frame of animation. The keys in the Time Slider are displayed as red lines.

Figure 1-22 *The* **Panels** *window*

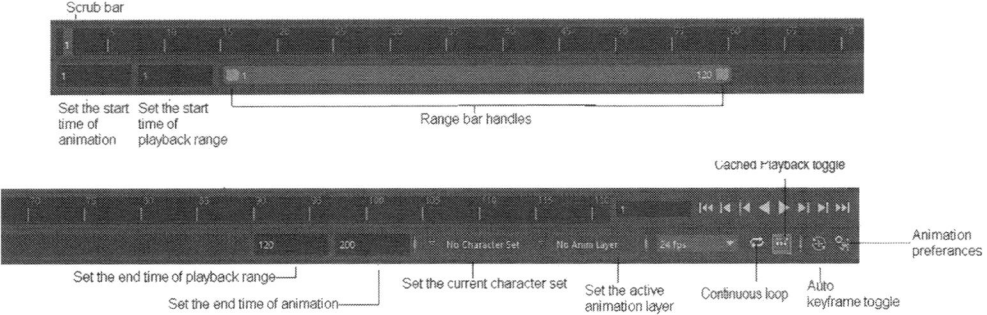

Figure 1-23 *The Time Slider and the Range Slider*

The Time Slider displays the range of frames available in your animation. In the Time Slider, the grey box, known as scrub bar, is used to move back and forth in the active range of frames available for animation. The Playback Controls at the extreme right of the current frame help you to play and stop the animation. The Range Slider located below the Time Slider is used to adjust the range of animation playback. The Range Slider shows the start and end time of the active animation. The edit boxes both on the left and right of the Range Slider direct you to the start and end frames of the selected range. The length of the Range Slider can be altered using these edit boxes. At the right of the **Set the end time of the animation** input box is the **Set the active animation layer** button. This feature gives you access to all the options needed to create and manipulate the animation layers. This option helps you to blend multiple animations in a scene. The **Set the current character set** is located on the right of the Range Slider. It is used to gain automatic control over the character animated object. There are two buttons on the extreme right of the Range Slider: **Auto keyframe toggle** and **Animation preferences**. These buttons are discussed next.

> **Tip**
> *You can also set the keys for animation by choosing* **Key > Set > Set key** *from the menubar or by pressing the 's' key. Ensure that you have selected the* **Animation** *menuset.*

Auto keyframe toggle

The **Auto keyframe toggle** button is used to set the keyframes. This button sets the keyframe automatically whenever an animated value is changed. Its color turns blue when it is activated.

Animation preferences

The **Animation preferences** button is used to modify the animation controls. On choosing this button, the **Preferences** window will be displayed, as shown in Figure 1-24. In the **Preferences** window, the **Time Slider** option is selected by default in the **Categories** area. You can set the animation controls in the **Time Slider** and **Playback** area of the **Preferences** window. Choose the **Save** button to save the changes and close the window.

Exploring Maya Interface

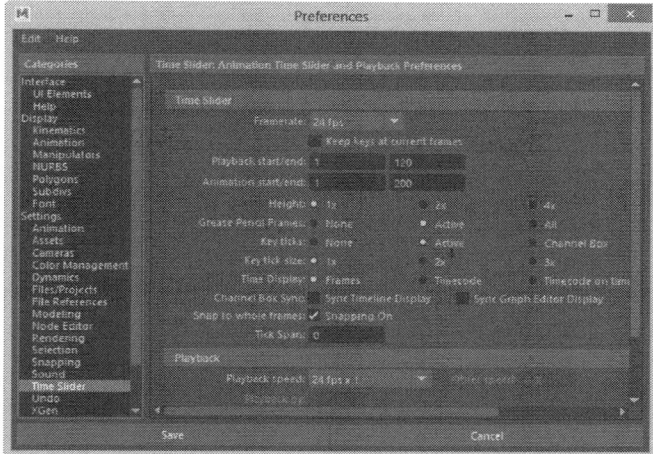

Figure 1-24 The **Preferences** *window*

Cached Playback

 Cached playback is the process that continuously evaluate the animation and helps to speed up the animation playback in the viewport. This appears as a blue stripe running along the bottom of the Time Slider. The **Cached Playback** button is used to change the animation without the need to create a playblast.

Command Line

The Command Line is located below the Range Slider. It works in Maya interface by using the MEL script or the Python script. The MEL and Python are the scripting languages used in Maya. Choose the **MEL** button to switch between the two scripts. The **MEL** button is located above the Help Line.

The Command Line also displays messages from the program in a grey box on the right. At the extreme right of the Command Line, there is an icon for the **Script Editor**. The **Script Editor** is used to enter complex and complicated MEL and Python scripts into the scene.

Note
MEL stands for MAYA Embedded Language. The **MEL** *command is a group of text strings that are used to perform various functions in Maya.*

Help Line

The Help Line is located at the bottom of the Command Line. It provides a brief description about the selected tool or the active area in the Maya interface.

Panel Menu

The **Panel** menu is available in every viewport, as shown in Figure 1-25. The commands or options in the **Panel** menu controls all the actions performed in the workspace. The **Panel** menu comprises of six menus, which are discussed next.

Figure 1-25 The **Panel** menu

Panel Toolbar

The **Panel** toolbar, as shown in Figure 1-26, is located just below the **Panel** menu of all viewports. This toolbar consists of the most commonly used tools present in the **Panel** menu. These tools are discussed next.

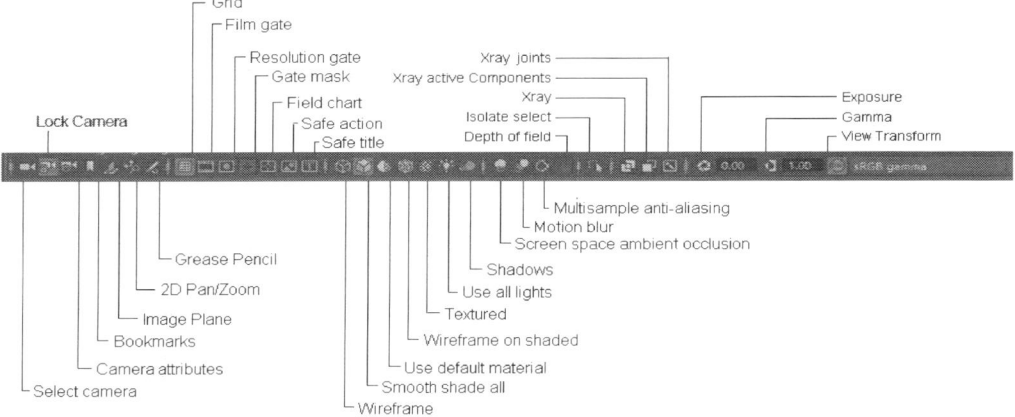

Figure 1-26 The **Panel** toolbar

Channel Box / Layer Editor

The **Channel Box** and the **Layer Editor** are used to edit the attributes of an object. The **Channel Box** consists of all object attributes used for editing, and the **Layer Editor** is used for creating layers for objects in the scene. To display the **Channel Box / Layer Editor**, choose **Windows > Editors > General Editors > Channel Box / Layer Editor** from the menubar. Alternatively, press the CTRL +A keys to open the **Channel Box / Layer Editor**, if it is not already displayed. Select an object; the attributes of the selected object will be displayed in the **Channel Box / Layer Editor**, refer to Figure 1-27. The **Channel Box** is further divided into three parts, which are discussed next.

Exploring Maya Interface

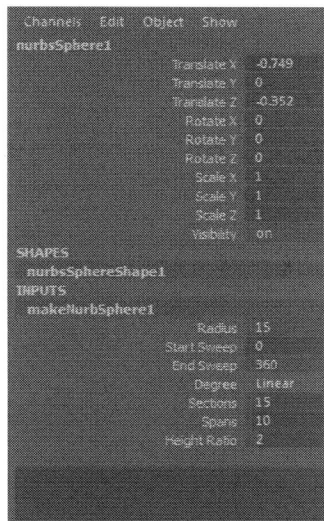

Figure 1-27 *The Channel Box /Layer Editor*

HOTKEYS

In Maya, you can create your own shortcut keys or even change default shortcuts. To do so, choose **Windows > Editors > Settings/Preferences > Hotkey Editor** from the menubar; the **Hotkey Editor** will be displayed, as shown in Figure 1-28. To edit hotkeys, select a hotkey category from the **Edit Hotkeys For** drop-down list. Now, find the desired command from the list displayed below the **Edit Hotkeys For** drop-down list. Click on the command and then enter a keyboard shortcut.

You can search an application command by choosing the **Search By** text box. Enter the application command name in the search bar; filtered items will be displayed as shown in Figure 1-29. At the right side of the **Hotkey Editor**, the **Keyboard** tab will be displayed. In this tab the unassigned keys are highlighted in cyan color.

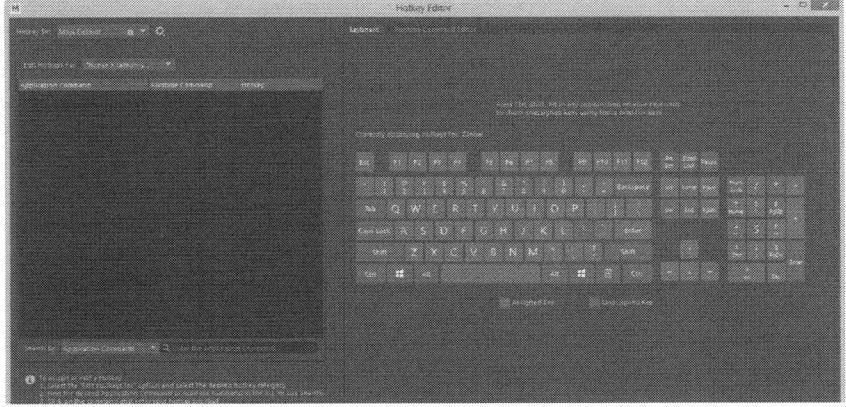

Figure 1-28 *The Hotkey Editor*

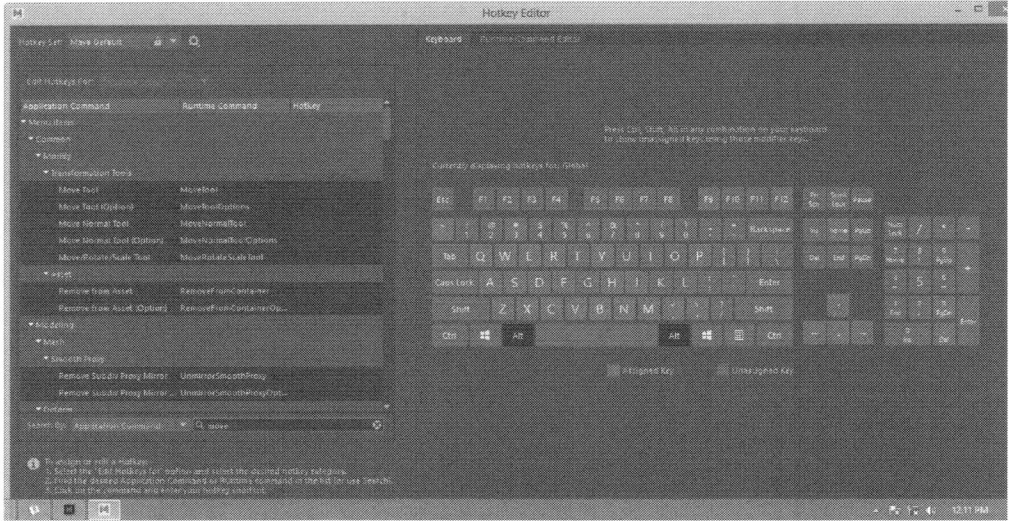

Figure 1-29 Using the **Search By** filter

HOTBOX

Hotbox, as shown in Figure 1-30, helps you access menu items in a viewport. The Hotbox is very useful, when you work in the expert mode or the full screen mode. It helps you access the menu items and tools by using cursor in the workspace. To access a command, press and hold the SPACEBAR key; the Hotbox will be displayed. Now, you can choose the option that you need to work from the Hotbox. The Hotbox is divided into five distinct zones, East, West, North, South, and Center, refer to Figure 1-30.

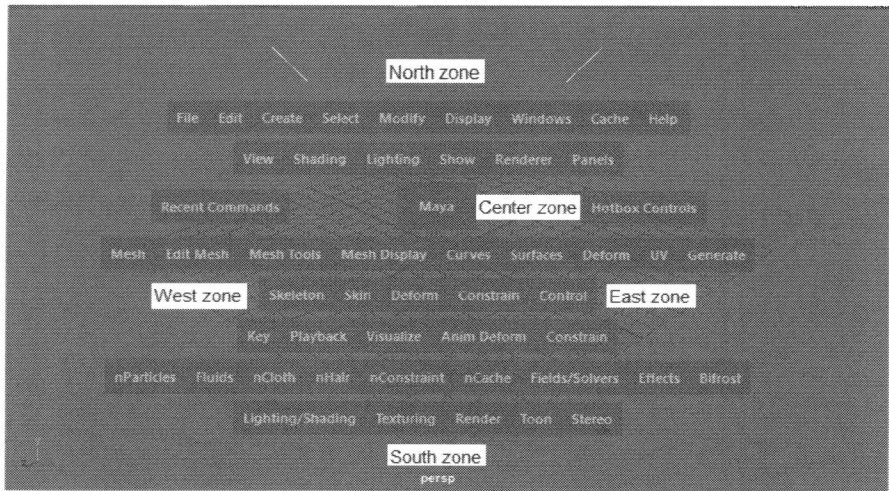

Figure 1-30 The Hotbox

Exploring Maya Interface

Note

You can turn off various UI elements in the Maya interface to get more space and then use the Hotbox to access various commands and tools. But you should do it only after you have established a workflow for yourself. In the beginning, you should use the menubar at the top of the screen instead of using the Hotbox as it reduces the possibility of confusion in finding a command at a later stage

OUTLINER

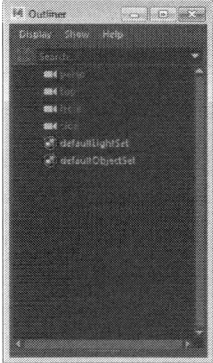

The **Outliner** window is used to display all the objects of a scene in a hierarchical manner, as shown in Figure 1-31. An object in the scene can be selected by simply clicking on its name in the **Outliner** window. In the **Outliner** window, the objects are placed in the order of their creation in the viewport. For example, if you create a cube in the viewport followed by a sphere and a cylinder, then all these objects will be placed in a sequential manner in the **Outliner** window, which means the object (cube) created first will be placed first and the object created last (cylinder) will be placed at the last. To organize the sequence manually, choose the MMB and then drag and drop one object below another object. To rename an object, double-click on the name of the object. At the top of the **Outliner** window, there is an text box known as the **Text Filter Box**. You can use this box to select objects with a particular name. For example, enter

Figure 1-31 Objects displayed in the Outliner window

front in the box and press ENTER; all the objects having the word 'front' in their name will be selected in the viewport. By default, there are four cameras in the **Outliner** window that represent four default viewports in Maya. As discussed earlier, everything that you see in the viewport is seen through the camera view. These cameras are visible in the **Outliner** window by default. Each object in the **Outliner** window has an icon of its own. When you double-click on any of these icons, the **Attribute Editor** will be displayed, where you can change the properties of various objects.

MARKING MENUS

Marking menus are similar to shortcut menus that consist of almost all the tools required to perform an operation on an object. There are three types of marking menus in Maya. The first type of marking menu is used to create default objects in the viewport. To create a default object, press and hold the SHIFT key and then right-click anywhere in the viewport; a marking menu will be displayed, as shown in Figure 1-32. In this marking menu, choose the object that you want to create.

The second type of marking menu is used to switch amongst various components of an object such as vertices, faces, edges, and so on. To invoke this marking menu, select an object and right-click; the marking menu will be displayed, as shown in Figure 1-33. Now, you can select the desired component of the selected object. This marking menu can also be used to apply material to an object. To do so, choose the **Assign New Material** option from this marking menu; the **Assign New Material** window will be displayed. Next, choose the required material; the material will be applied to the selected object. This method will be discussed in detail in later chapters. The third type of marking menu is used to modify the components of an object. To invoke this marking menu, select a component, press and hold the SHIFT key, and then right-click on the selected object; the marking menu will be displayed, refer to Figure 1-33. After invoking this marking menu, you can choose the desired option to perform the corresponding function.

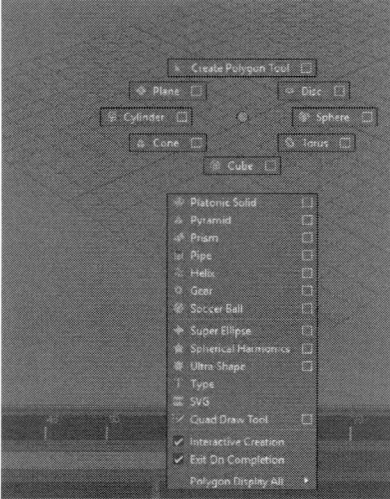

Figure 1-32 Marking menu displaying options used for creating default objects

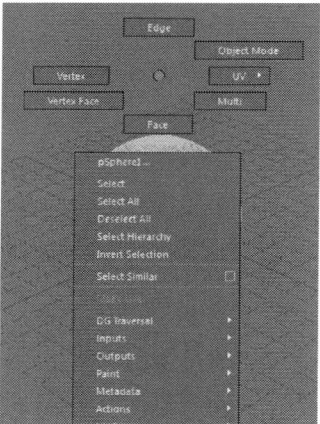

Figure 1-33 Marking menu displaying components of the selected object

PIPELINE CACHING

In Maya, you can reduce the render time of a complex scene with the help of pipeline cache tools. Using these tools, you can also increase the loading speed of large 3D scenes. The two types of caching tools available in Maya are discussed next.

Alembic Cache

The alembic cache enables you to save and export complex Maya scenes in alembic file format. The alembic file format has been developed to represent a complex 3D geometry as a simple geometry. The exported alembic files can then be re-imported into Maya to improve playback performance and reduce memory usage. In order to access this tool, choose **Cache > Alembic Cache** from the menubar; a flyout will be displayed, as shown in Figure 1-34. Various options in this flyout are discussed next.

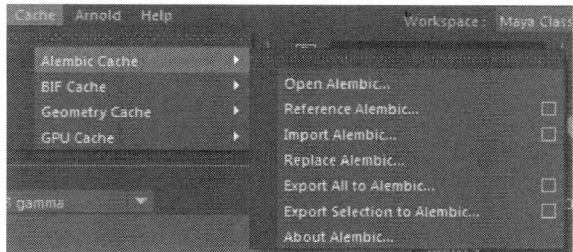

*Figure 1-34 Flyout displayed on choosing the **Alembic Cache** tool from the menubar*

Note
*The **Alembic Cache** tool is not available by default. To invoke this tool, run the following script in Command Line: global proc perFrameCallback(int $frame){print $frame;}*

INTEROPERABILITY OPTIONS IN MAYA

Autodesk Maya enables you to exchange data between Maya and different softwares such as 3ds Max, Unity, and Print Studio. However, for exchanging data, the same version of the software must be available on your system. The **Send to 3ds Max**, **Send to Unity**, and **Send to Print Studio** options located in the **File** menu of the menubar are used to send a Maya file to any of the above mentioned software.

Note
*The **Send to 3ds Max** option located in the **File** menu of the menubar will be displayed only if you have matching versions installed on your system. For example, 3ds Max 2019 and Maya 2019 are considered to be the matching versions.*

NAVIGATING THE VIEWPORTS

The persp view is the default camera view in Maya. To look around in a scene, you can move the virtual camera associated with the viewport. You can use the following shortcut keys while navigating the viewport.

Keyboard Shortcut	Function
ALT+MMB+Drag	Helps to pan the viewport
ALT+RMB+Drag	Helps to dolly in and out the viewport. You can also use the scroll wheel to dolly in and out.
ALT +LMB+Drag	Rotates or orbits the camera around in the persp window

WORKSPACES

In this version of Maya, workspaces are introduced. Workspaces are arrangement of windows, panels, and other interface elements. Maya comes with several predefined workspaces that you can access from the **Workspaces** drop-down list available on the far right of the menubar, as shown in Figure 1-35.

You can also save your own workspaces or reset the factory workspaces. To do so, choose the options available in the **Windows > Workspaces** menu.

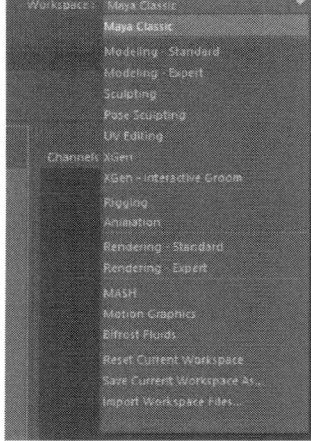

Figure 1-35 Partial view of the *Workspace* drop-down list

Chapter 2

Polygon Modeling

Learning Objectives
After completing this chapter, you will be able to:
- *Create polygon primitives*
- *Edit polygon primitives*
- *Create models using polygon primitives*

INTRODUCTION

In this chapter, you will learn to create and edit polygon shapes using polygon modeling techniques. A polygon is made up of different closed planar shapes having straight sides. The most commonly used shapes in 3D polygons are triangles and quadrilaterals. These shapes are formed by vertices, edges, and faces. An edge is a straight line formed by joining two vertices. In a polygon, three vertices join to each other by three edges to form a triangle and four vertices join to each other by four edges to form a quadrilateral. By modifying faces, edges, and vertices of an object, you can create a polygon model as per your requirement.

Note

*If you want to create polygon objects using click-drag operations, you need to turn on the **Interactive Creation** option available in the menubar. To do so, choose **Create > Objects > Polygon Primitives > Interactive Creation** from the menubar. The **Interactive Creation** option works with all primitives. There are certain parameters that cannot be controlled via interactive creation. These parameters can only be changed from the settings window of the tool.*

*This option also affects how Maya shows the tool settings. For example, if the **Interactive Creation** option is selected and you choose **Create > Objects > Polygon Primitives > Sphere > Option Box** from the menubar, the **Tool Settings (Polygon Sphere Tool)** panel will be displayed. In this panel, you can set non-interactive attributes such as **Axis divisions** and **Height divisions** and then click-drag in the viewport to interactively define the radius of the sphere. If you want to create a sphere with the current settings specified in the panel, just click on the viewport instead of clicking and dragging. You can reset the settings by choosing the **Reset Tool** button available at the top-right corner of the panel.*

*If the **Interactive Creation** option is not selected, the **Polygon Sphere Options** window will be displayed. In this window, specify the attributes and then choose the **Create** button to create sphere with specified settings.*

POLYGON PRIMITIVES

In Maya, polygon primitives are classified into various objects. These objects are grouped under **Polygon Primitives** in the menubar.

Creating a Sphere

Menubar:	Create > Objects > Polygon Primitives > Sphere
Shelf:	Polygons > Polygon Sphere

A sphere is a solid object in which every point on its surface is equidistant from its center, as shown in Figure 2-1. The sphere can be created interactively or by entering the values using the keyboard. Both the methods are discussed next.

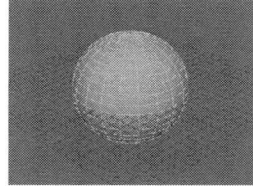

Figure 2-1 A polygon sphere

EDITING THE POLYGON COMPONENTS

In the previous section, you learned to modify simple polygon primitives. In this section, you will learn to edit the components of polygon primitives to create complex objects from it. To do so, select a polygon object in the viewport and then press and hold the right mouse button over it; the marking menu of the corresponding object will display various components of the object such as vertex, edge, face, and UV, refer to Figures 2-2 to 2-5. To access various tools for editing the polygon primitives, select **Modeling** from the **Menuset** drop-down list in Status Line. Next, choose the **Edit Mesh** menu from the menubar. The most commonly used component editing tools are discussed next.

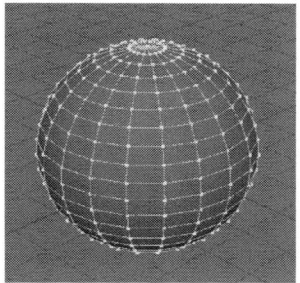 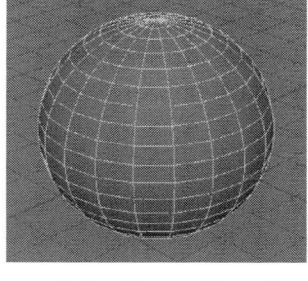

Figure 2-2 Vertices of the sphere   *Figure 2-3* Edges of the sphere

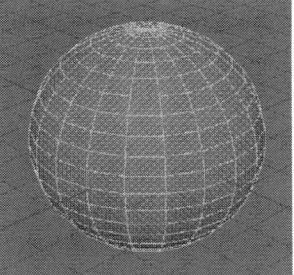 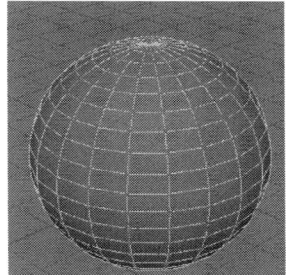

Figure 2-4 Faces of the sphere   *Figure 2-5* UVs of the sphere

TUTORIAL

All the files used in the tutorials can be downloaded from the CADSoft website (*www.cadsofttech.com*). These files are compressed in zip file format and are required to be extracted before using them in the tutorials. The path of the files is as follows: *Textbooks > Animation and Visual Effects > Maya > Autodesk Maya 2019 for Novices*

Tutorial 1

In this tutorial, you will create model of a coffee mug shown in Figure 2-6 using the polygon modeling techniques.   (**Expected time: 20 min**)

Figure 2-6 The model of a coffee mug

The following steps are required to complete this tutorial:

a. Create a project folder.
b. Create the basic shape of the mug.
c. Create the handle of the mug.
d. Change the background color of the scene.
e. Save and render the scene.

Creating a Project Folder

Before starting a new scene, it is recommended that you create a project folder. It helps you keep all the files of a project in an organized manner. Open Windows Explorer and browse to the *Documents* folder. In this folder, create a new folder with the name *maya2019*. The *maya2019* folder will be the main folder and it will contain all the projects folders that you will create while doing tutorials of this textbook. Now, you will create a project folder for Tutorial 1 of this chapter. To do so, you need to follow the steps given next.

1. Start Autodesk Maya 2019 by double-clicking on its icon on the desktop.

2. Choose **File > Project > Project Window** from the menubar; the **Project Window** is displayed. Choose the **New** button; the **Current Project** and **Location** text boxes are enabled. Now, enter **c02_tut1** in the **Current Project** text box.

3. Click on the folder icon next to the **Location** text box; the **Select Location** dialog box is displayed. In this dialog box, browse to the *\Documents\maya2019* folder and choose the **Select** button to close the dialog box. Next, choose the **Accept** button in the **Project Window** dialog box; the *\Documents\maya2019\c02_tut1* folder will become the current project folder.

4. Choose **Save Scene** from the **File** menu; the **Save File As** dialog box is displayed.

Note
*The scenes created in Maya are saved with the .ma or .mb extension. As the project folder is already created, the path \Documents\maya2019\c02_tut1\scenes is displayed in the **Look in** drop-down list of the **Save As** dialog box.*

Tip
After setting the project folder, when you open or save a scene, Maya uses the scenes folder inside the project folder by default.

Polygon Modeling

5. Enter **c02tut1** in the **File name** edit box and then choose the **Save As** button to close the dialog box.

Note
It is recommended that you frequently save the file while you are working on it by pressing the CTRL+S keys.

Creating the Basic Shape of the Mug

In this section, you will use the **Cylinder** polygon primitive to create the basic shape of the mug.

1. Choose **Create > Objects > Polygon Primitives > Cylinder > Option Box** from the menubar; the **Tool Settings (Polygon Cylinder Tool)** panel is displayed in the viewport. Enter the values in the **Tool Settings (Polygon Cylinder Tool)** panel, as shown in Figure 2-7.

2. Click in the persp viewport; a cylinder is created, refer to Figure 2-8. Close the **Tool Settings (Polygon Cylinder Tool)** panel.

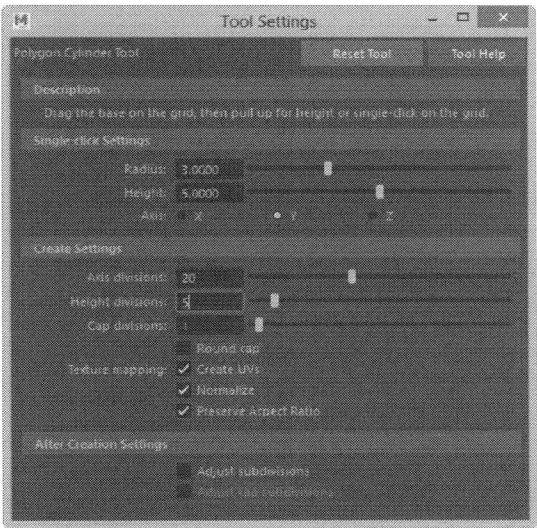

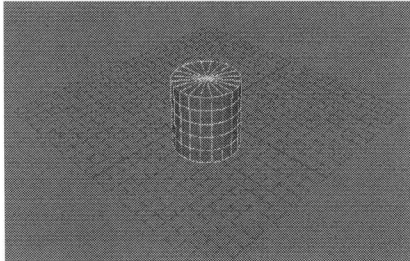

*Figure 2-7 The **Tool Settings (Polygon Cylinder Tool)** panel*

Figure 2-8 Cylinder created in the viewport

3. In the **Channel Box / Layer Editor**, click on the **pCylinder1** tab; a text box is activated. Next, type **mug** in the text box and press ENTER; the **pCylinder1** tab is renamed as *mug*.

4. Hover the cursor in the persp viewport and press SPACEBAR; the four viewports are displayed. Next, hover the cursor on the front-Z viewport and press SPACEBAR; the front-Z viewport is maximized.

   Select *mug* if not selected and then press and hold the right mouse button; a marking menu is displayed.

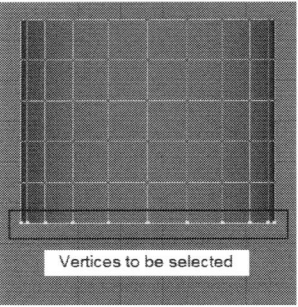

Figure 2-9 Bottom vertices of the cylinder selected

5. Choose **Vertex** from the marking menu; the vertex selection mode is activated.

6. Select the vertices at the bottom of *mug*, refer to Figure 2-9. Next, invoke the **Scale Tool** by pressing the R key.

7. Scale down the selected vertices of *mug* inward uniformly, as shown in Figure 2-10. Similarly, select the other loops of vertices and scale them to form the shape of a mug, refer to Figure 2-11.

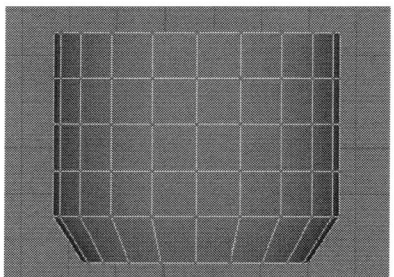 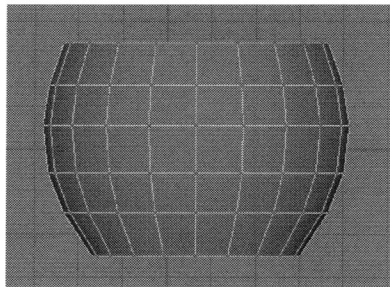

Figure 2-10 Bottom vertices of the cylinder scaled

Figure 2-11 Basic shape of the mug created

Next, you need to add segments at the top and bottom.

8. Make sure the **Modeling** menuset is selected in the **Menuset** drop-down list. Choose **Mesh Tools > Tools > Insert Edge Loop** from the menubar. Next, click at the top and bottom region of *mug*; two edges are inserted, refer to Figure 2-12. Deactivate the **Insert Edge Loop** tool by pressing the W key.

Polygon Modeling

9. Maximize the persp viewport. Make sure *mug* is selected and then press and hold the right mouse button; a marking menu is displayed. Choose **Face** from the marking menu; the face selection mode is activated. Now, select the top faces of *mug* using the SHIFT key, refer to Figure 2-13. Next, choose **Edit Mesh > Components > Extrude** from the menubar.

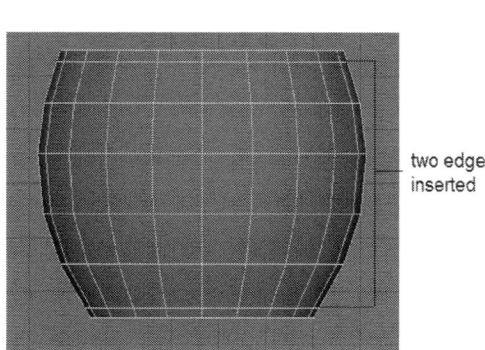

Figure 2-12 Two edges inserted at the top and bottom of the cylinder

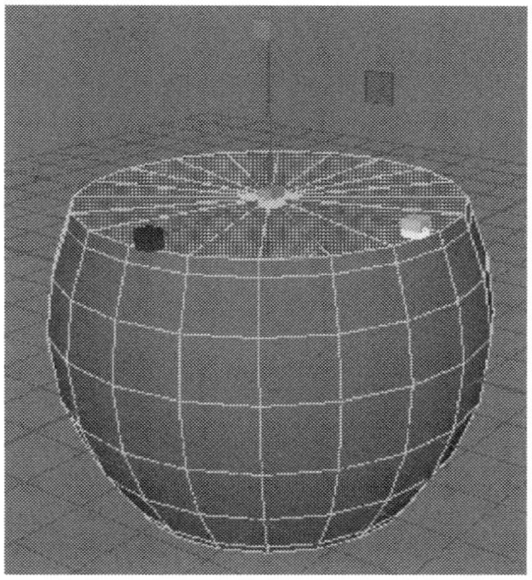

Figure 2-13 Top faces of the cylinder selected

10. Invoke the **Scale Tool** and scale down the selected faces uniformly, refer to Figure 2-14.

11. Again, choose **Edit Mesh > Components > Extrude** from the menubar; the **polyExtrudeFace#** In-View Editor is displayed in the viewport, refer to Figure 2-15. Enter **-0.3** in the **Thickness** edit box of the **polyExtrudeFace#** In-View Editor, refer to Figure 2-15; the shaded faces are extruded.

12. Press the G key to invoke the **Extrude** tool again and enter the value **-1.6** in the **Thickness** edit box; the top faces of *mug* are extruded downward.

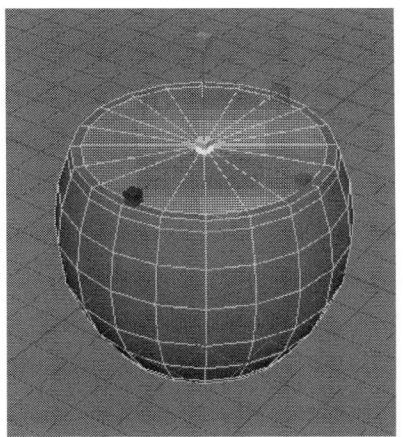
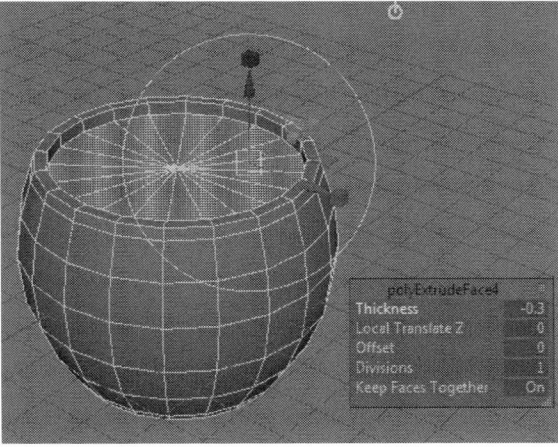

Figure 2-14 The top selected faces of the mug scaled down using the **Scale Tool**

Figure 2-15 The **polyExtrudeFace#** In-View Editor displayed

Note

The G key is used to repeat the last performed action in Maya.

13. Press G again to invoke the **Extrude** tool, and enter the value **-2** in the **Thickness** edit box. Next, enter **0.8** in the **Offset** edit box; the selected polygon is extruded inward. Deactivate the **Extrude** tool.

14. Maximize the top-Y viewport such that you can view the inner area of *mug*. Press 3 to view the object in the smooth mode. To rectify the distortion in the geometry, you need to add edges. Press 1 and choose **Mesh Tools > Tools > Insert Edge Loop** tool; the shape of the cursor changes and then insert two edges inside the mug, refer to Figure 2-16. Deactivate the **Insert Edge Loop** tool by pressing W.

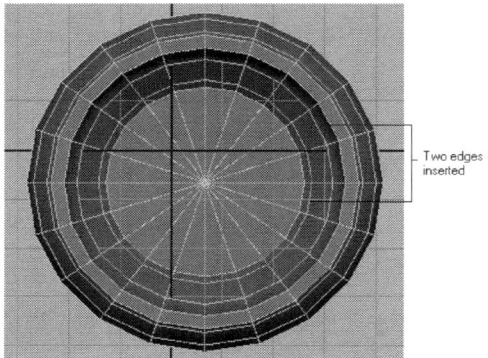

Figure 2-16 *Two edge loops added inside the mug*

Creating the Handle of the Mug

In this section, you need to create the handle of the mug.

1. Maximize the side-X viewport. Move the cursor over *mug* and then press and hold the right mouse button; a marking menu is displayed. Choose **Edge** from the marking menu; the edge selection mode is activated.

2. Select two edges of *mug*, refer to Figure 2-17. Next, choose **Edit Mesh > Components > Bevel > Option Box**; the **Bevel Options** window is displayed. Now, enter **1** in the **Width** edit box and choose the **Bevel** button; the selected edges will be beveled, as shown in Figure 2-18.

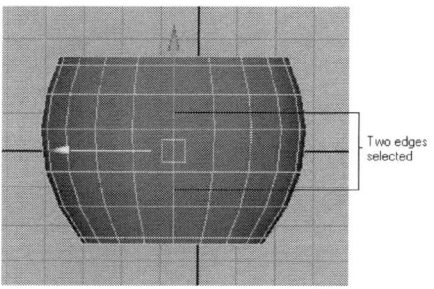
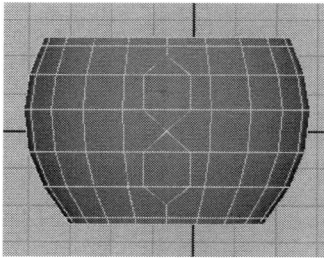

Figure 2-17 *Two edges of mug selected*   *Figure 2-18* *Selected edges beveled*

Polygon Modeling

3. Move the cursor over *mug* and then press and hold the right mouse button; a marking menu is displayed. Choose **Face** from the marking menu; the face selection mode is activated. Next, select a face of *mug*, as shown in Figure 2-19.

4. Choose **Edit Mesh > Components > Extrude** from the menubar. Next, invoke the **Scale Tool** by pressing the R key and scale down the selected face of *mug* uniformly upto 70%. You can check the scale size in the status line, as shown in Figure 2-20.

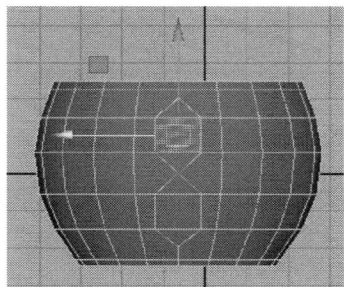 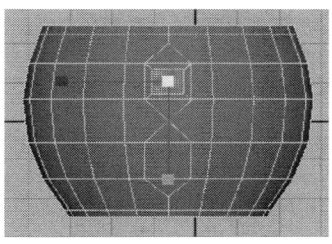

Figure 2-19 A face of mug selected   *Figure 2-20* Face of the mug scaled down

5. Select the face of *mug*, as shown in Figure 2-21. Repeat the procedure as done in Step 4 to scale down the face, refer to Figure 2-22.

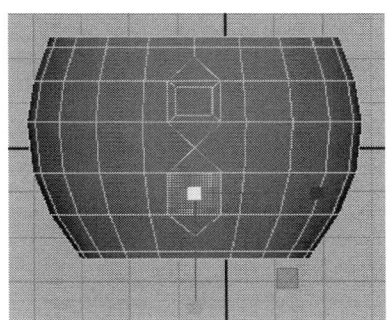 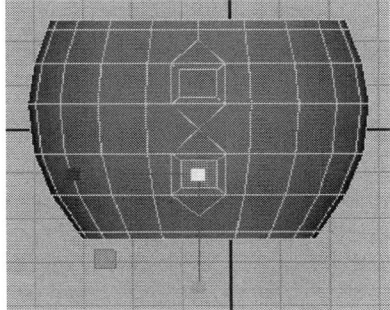

Figure 2-21 A face of the mug selected   *Figure 2-22* A face of the mug scaled down

6. Maximize the persp viewport. Make sure that both the scaled faces are selected, and then invoke the **Extrude** tool by pressing the G key. Next, enter the value **0.8** in the **Thickness** edit box of the **polyExtrudeFace#** In-View Editor.

7. Deactivate the **Extrude** tool by pressing the W key. Make sure the two extruded faces are selected. Next, choose **Edit Mesh > Components > Bridge > Option Box** from the menubar; the **Bridge Options** window is displayed. Enter values in the **Bridge Options** window, as shown in Figure 2-23. Next, choose the **Apply** button and close the window; the extruded faces are connected to each other.

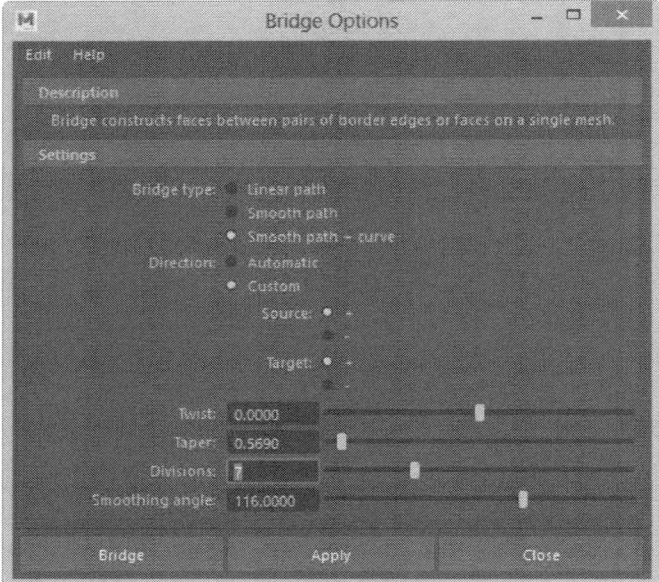

*Figure 2-23 The **Bridge Options** window*

8. Make sure *mug* is selected and then press and hold the right mouse button on it; a marking menu is displayed. Next, choose **Object Mode** from the marking menu; the object selection mode is activated.

9. Select *mug* and then choose **Mesh > Remesh > Smooth** from the menubar; the mesh of *mug* is smoothened. Press SPACEBAR; the four viewports display the *mug* after applying **Smooth Tool**, as shown in Figure 2-24.

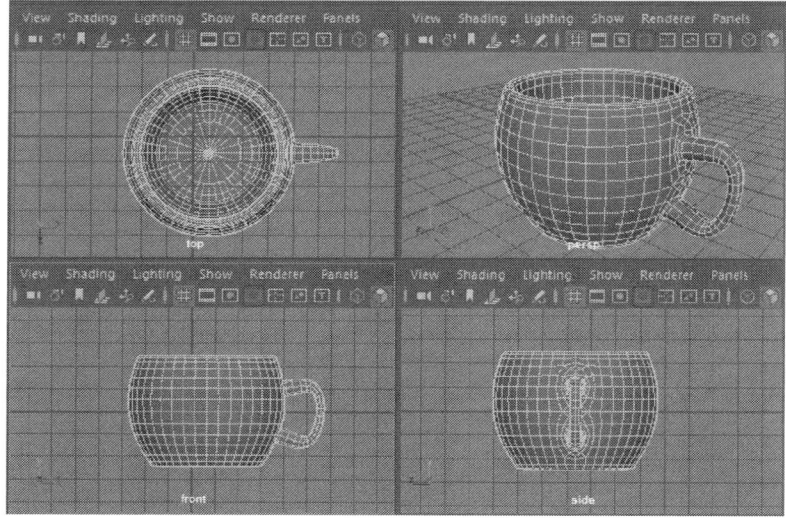

Figure 2-24 The mug displayed in all viewports

Polygon Modeling

Changing the Background Color of the Scene

In this section, you will change the background color of the scene.

1. Choose **Windows > Editors > Outliner** from the menubar; the **Outliner** window is displayed. Select the **persp** camera in the **Outliner** window; the **perspShape** tab is displayed in **Attribute Editor**.

2. In the **perspShape** tab, expand the **Environment** node and drag the **Background Color** slider bar toward right to change the background color to white.

Saving and Rendering the Scene

In this section, you will save the scene that you have created and then render it. You can view the final rendered image of the scene by downloading the *c02_maya_2019_rndr.zip* file from *www.cadsofttech.com*. The path of the file is as follows: *Textbooks > Animation and Visual Effects > Maya > Autodesk Maya 2019 for Novices*.

1. Choose **File > Save Scene** from the menubar.

2. Maximize the persp viewport if not already maximized. Choose the **Display render setting** button from the Status Line; the **Render Settings** window is displayed. In this window, change renderer to **Maya Software** and then close the window. Choose the **Render the current frame** button from the Status Line to render the scene.

EXERCISES

The rendered output of the models used in the following exercises can be accessed by downloading the file *c02_maya_2019_exr.zip* from *www.cadsofttechcom*. The path of the file is as follows: *Textbooks > Animation and Visual Effects > Maya > Autodesk Maya 2019 for Novices*

Exercise 1

Using various polygon modeling techniques, create the model of a USB cable, as shown in Figure 2-25. (**Expected time: 30 min**)

Figure 2-25 Model to be created in Exercise 1

Exercise 2

Using various polygon modeling techniques, create a scene, as shown in Figure 2-26.

(Expected time: 30 min)

Figure 2-26 Scene to be created in Exercise 2

Chapter 3

NURBS Curves and Surfaces

Learning Objectives

After completing this chapter, you will be able to:
- *Create NURBS Primitives*
- *Create NURBS curves*
- *Create surfaces*

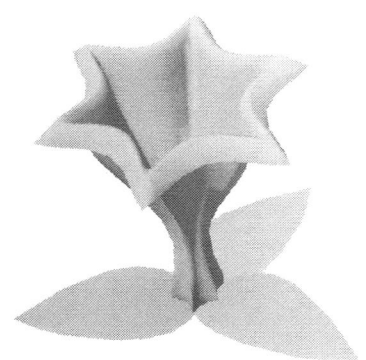

INTRODUCTION

In Maya, there are three different types of modeling: NURBS, polygon, and subdivision surface. NURBS, which stands for Non-Uniform Rational B-Splines, is an industry standard for designing and modeling surfaces. NURBS modeling is suitable for modeling surfaces with complex curves. NURBS surfaces can be manipulated interactively with ease. Before modeling an object, you need to visualize it in 3D terms. Visualization of an object in 3D terms helps you in determining the type of modeling that you need to use for creating the object. In this chapter, you will learn about various NURBS modeling tools and techniques.

NURBS PRIMITIVES

In this chapter, you will learn about NURBS curves and surfaces. NURBS (Non-Uniform Rational B-Spline) is a mathematical representation of 3D geometry that can describe any shape accurately. NURBS modeling is basically used for creating curved shapes and lines.

In Maya, there are default NURBS objects that resemble various geometrical objects. These NURBS objects are grouped together under the NURBS Primitives group in the menubar. To access the NURBS primitives, choose **Create > Objects > NURBS Primitives** from the menubar; a cascading menu will be displayed with all the default NURBS primitives. Some of the NURBS primitives can also be accessed from the Shelf, refer to Figure 3-1. In order to access the NURBS modeling tools for the NURBS primitives, make sure that the **Curves/Surfaces** Shelf tab is chosen from the Shelf. The types of NURBS Primitives are discussed next.

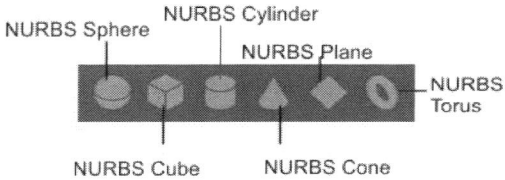

Figure 3-1 Accessing NURBS primitives from the Shelf

Note
*To create a NURBS object dynamically, you need to turn on the **Interactive Creation** option. To do so, choose **Create > Objects > NURBS Primitives > Interactive Creation** from the menubar.*

Creating a Sphere

Menubar:	Create > Objects > NURBS Primitives > Sphere
Shelf:	Curves/Surfaces > NURBS Sphere

A sphere is a solid object and every point on its surface is equidistant from its center, as shown in Figure 3-2. To create a sphere, choose **Create > Objects > NURBS Primitives > Sphere** from the menubar; a sphere will be created in all viewports. Alternatively, choose the **NURBS Sphere** tool from the **Curves/Surfaces** Shelf tab. You can create a sphere either dynamically or by entering values using the keyboard. Both the methods are discussed next.

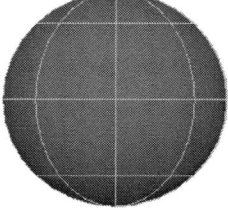

Figure 3-2 The NURBS sphere

NURBS Curves and Surfaces

Creating a Cube

Menubar: Create > Objects > NURBS Primitives > Cube
Shelf: Curves/Surfaces > NURBS Cube

A cube is a three-dimensional shape with six sides, as shown in Figure 3-3. To create a **NURBS cube**, choose **Create > Objects > NURBS Primitives > Cube** from the menubar; the sphere will be created in all viewports. Alternatively, to create a cube, you can choose the **NURBS Cube** tool from the **Curves/Surfaces** Shelf tab. You can also create a cube dynamically or by entering values using the keyboard. Both the methods are discussed next.

Figure 3-3 The NURBS cube

Creating a Cylinder

Menubar: Create > Objects > NURBS Primitives > Cylinder
Shelf: Curves/Surfaces > NURBS Cylinder

A cylinder is a solid geometry with straight parallel sides and circular sections, as shown in Figure 3-4. To create a cylinder in the viewport, choose **Create > Objects > NURBS Primitives > Cylinder** from the menubar; the cylinder will be created in all viewports. Alternatively, you can choose the **NURBS Cylinder** tool from the **Curves/Surfaces** Shelf tab. You can create a cylinder either dynamically or by entering values using the keyboard. Both the methods of creating the cylinder are discussed next.

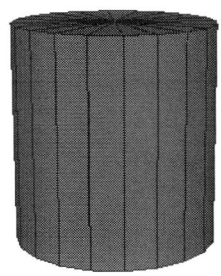

Figure 3-4 The NURBS cylinder

Creating a Cone

Menubar: Create > Objects >NURBS Primitives > Cone
Shelf: Curves/Surfaces > NURBS Cone

A cone is an object with a circular base and its sides tapered up to a point, as shown in Figure 3-5. To create a cone, choose **Create > Objects > NURBS Primitives > Cone** from the menubar; the cone will be created in all viewports. Alternatively, you can create a cone by invoking the **NURBS Cone** tool from the **Curves/Surfaces** Shelf tab. You can create a cone either dynamically or by entering values using the keyboard. Both the methods of creating a cone are discussed next.

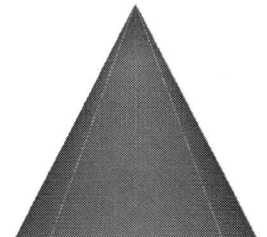

Figure 3-5 The NURBS cone

Creating a Plane

Menubar: Create > Objects > NURBS Primitives > Plane
Shelf: Curves/Surfaces > NURBS Plane

A plane is a two-dimensional flat surface, as shown in Figure 3-6. To create a NURBS plane, choose **Create > Objects > NURBS Primitives > Plane** from the menubar. Alternatively, you can create a plane by invoking the **NURBS Plane** tool from the **Curves/Surfaces** Shelf tab. You can create a plane either dynamically or by entering values using the keyboard. Both the methods of creating a plane are discussed next.

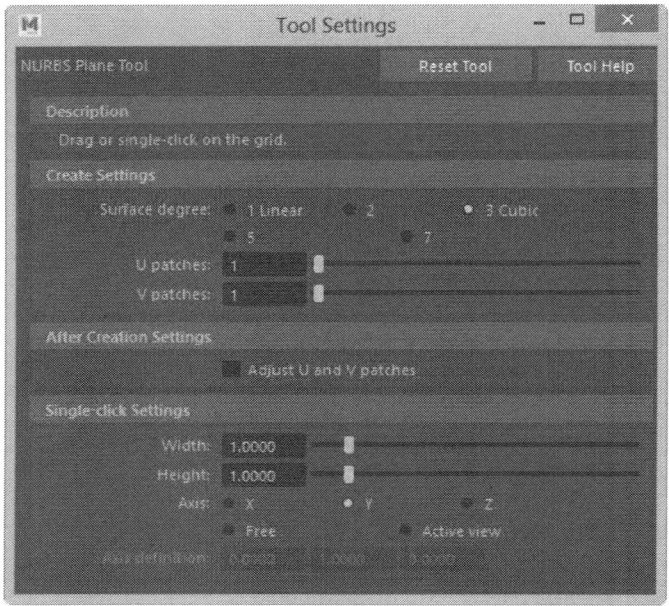

Figure 3-6 The Tool Settings (NURBS Plane Tool) panel

Creating a Torus

Menubar: Create > Objects > NURBS Primitives > Torus
Shelf: Curves/Surfaces > NURBS Torus

A torus is created by revolving a circular profile around a circular or an elliptical path. To create a NURBS torus, choose **Create > Objects > NURBS Primitives > Torus** from the menubar. Alternatively, you can create a torus by choosing the **NURBS Torus** tool from the **Curves/Surfaces** Shelf tab. You can create a torus either dynamically or by entering values using the keyboard. Both the methods of creating a torus are discussed next.

NURBS Curves and Surfaces

Creating a Circle

Menubar: Create > Objects > NURBS Primitives > Circle
Shelf: Curves/Surfaces > NURBS Circle

A circle is a closed plane curve in which every point on the curve is equidistant from the center, as shown in Figure 3-7. To create a circle, choose **Create > Objects > NURBS Primitives > Circle** from the menubar. Alternatively, you can create a circle by choosing the **NURBS Circle** tool from the **Curves/Surfaces** Shelf tab. You can create a circle either dynamically or by entering values using the keyboard. Both the methods of creating circle are discussed next.

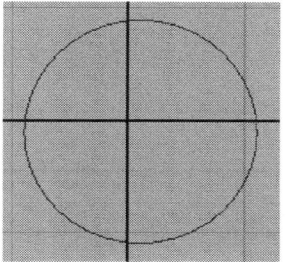

Figure 3-7 The NURBS circle

Creating a Square

Menubar: Create > Objects > NURBS Primitives > Square
Shelf: Curves/Surfaces > NURBS Square

A square is a four-sided regular polygon with equal sides, as shown in Figure 3-8. To create a square, choose **Create > Objects > NURBS Primitives > Square** from the menubar; the square will be created in all viewports. Alternatively, you can create a square by invoking the **NURBS Square** tool from the **Curves** Shelf tab. You can create a square either dynamically or by entering values by using the keyboard. Both the methods of creating a square are discussed next.

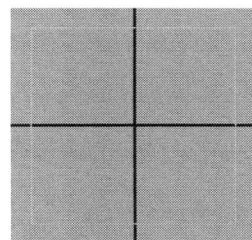

Figure 3-8 The NURBS square

WORKING WITH NURBS COMPONENTS

Each NURBS object has certain components such as **Isoparm**, **Hull**, **Surface Patch**, **Surface UV**, **Control Vertex**, and **Surface Point**, refer to Figures 3-9 to 3-14. To view the components of a NURBS object, select the NURBS object in the viewport and choose **Display > NURBS** from the menubar; a cascading menu will be displayed. Choose the component that you want to modify from the cascading menu; the selected component will be displayed in the viewport. Alternatively, press and hold the right mouse button over the object and choose the required component from the marking menu.

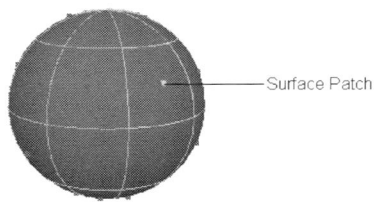

Figure 3-9 Surface Patch of the NURBS

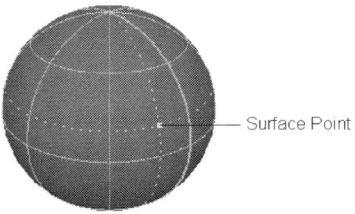

Figure 3-10 Surface Point of the NURBS

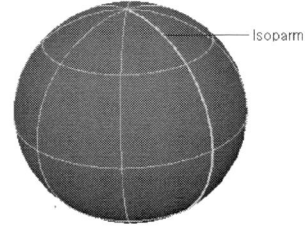

Figure 3-11 Isoparm of the NURBS

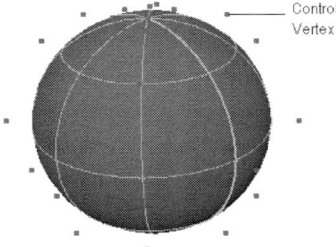

Figure 3-12 Control Vertex of the NURBS

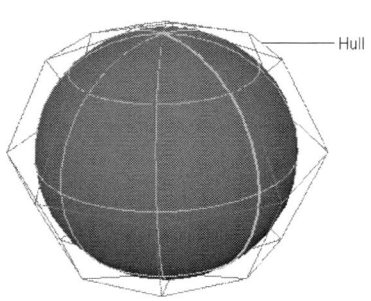

Figure 3-13 Hull of the NURBS

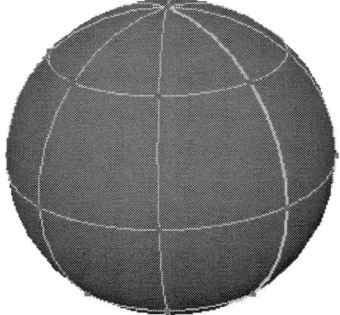

Figure 3-14 Surface UV of the NURBS

TOOLS FOR CREATING NURBS CURVES

In Maya, you can create NURBS curves using various tools. The tools used to create NURBS curves are discussed next.

CV Curve Tool

Menubar: Create > Objects > Curve Tool > CV Curve Tool

The **CV Curve Tool** is used to create curves in the viewport. A CV curve comprises of control vertices or CVs. To create a CV curve, choose **Create > Objects > Curve Tools > CV Curve Tool** from the menubar; the cursor will change into a plus sign. Next, click on different places in the viewport to create a curve. The first CV of the curve will be displayed as a box, and the second CV will be displayed as letter U. The box defines the starting point of the curve, and the letter U defines the direction of the curve. Press ENTER to finish the curve creation process. To edit the properties of a curve, choose **Create > Objects > Curve Tools > CV Curve Tool Option Box** from the menubar; the **Tool Settings (CV Curve Tool)** panel will be displayed, as shown in Figure 3-15. The options in the panel are discussed next.

Curve degree

The radio buttons corresponding to the **Curve degree** attribute are used to define the smoothness of a curve. By default, the **3 Cubic** radio button is selected in the **Curve degree** area. The higher the degree of curve, the smoother it will be.

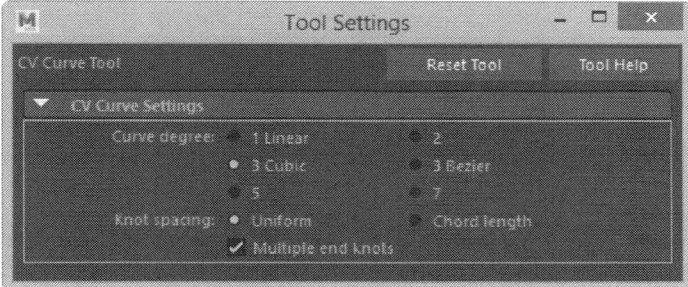

Figure 3-15 The *Tool Settings (CV Curve Tool)* panel

Knot spacing

The radio buttons corresponding to the **Knot spacing** attribute are used to define the distribution of the knots on the curve. Knots are the parametric locations(u) along the curve. The **Knot spacing** attribute has two radio buttons: **Uniform** and **Chord length**. The **Uniform** radio button is selected by default and is used to create the U parametric location values that are easier to predict. The **Chord length** radio button is used to distribute the curvature in such a way that the surface displays a symmetrical texture applied over it.

Tip
By default, CV Curve Tool is not present in the Curves Shelf tab. To add CV Curve Tool to the Shelf, press and hold CTRL+SHIFT and choose Create > CV Curve Tool from the menubar; CV Curve Tool icon will be displayed in the Shelf.

EP Curve Tool

Menubar:	Create > Objects > Curve Tools > EP Curve Tool
Shelf:	Curves/Surfaces > EP Curve Tool

The **EP Curve Tool** is also used to create an outline of a curve by placing edit points on it. To create an outline, choose **Create > Objects > Curve Tools > EP Curve Tool** from the menubar; the cursor sign will change into a plus sign. Now, click on different places in the viewport to create a curve. Next, press ENTER to finish the curve creation process. To modify the properties of the EP curve, choose **Create > Objects > Curve Tools > EP Curve Tool > Option Box** from the menubar; the **Tool Settings (EP Curve Tool)** panel will be displayed. Alternatively you can invoke this panel form the **Curve / Surfaces** Shelf tab by double-click on the icon in the Shelf tab; the **Tool Settings (EP Curve Tool)** panel will be displayed. The options in the **Tool Settings (EP Curve Tool)** panel are similar to those discussed in the **Tool Settings (CV Curve Tool)** panel.

Note
*The process of creating a curve using **EP Curve Tool** is different from that of **CV Curve Tool**. In both the cases, if **3 cubic** is selected from the **Curve degree** attribute, then the curve created using **CV Curve Tool** will create a smooth curve in the fourth segment whereas in case of **EP curve Tool**, a smooth curve will be created in the third segment.*

Pencil Curve Tool

Menubar:	Create > Objects > Curve Tools > Pencil Curve Tool
Shelf:	Curves/Surfaces > Pencil Curve Tool

The **Pencil Curve Tool** works similar to the brush tool available in other softwares. This tool is used to draw a freehand NURBS curve. To do so, choose **Create > Objects > Curve Tools > Pencil Curve Tool** from the menubar; the cursor will change into a pencil sign. Next, press and hold the left mouse button and drag the cursor in the viewport to create a curve. To set the properties of the curve, choose **Create > Objects > Curve Tools > Pencil Curve Tool > Option Box** from the menubar; the **Tool Settings (Pencil Curve Tool)** panel will be displayed. Alternatively you can invoke this panel form the **Curves/Surfaces** Shelf tab by double-click on the icon in the **Curves/Surfaces** Shelf tab; the **Tool Settings (Pencil Curve Tool)** panel will be displayed. The options in the **Tool Settings (Pencil Curve Tool)** panel are similar to those discussed in the **Tool Settings (EP Curve Tool)** panel.

Arc Tools

Menubar:	Create > Objects > Curve Tools
Shelf:	Curves/Surfaces > Three Point Circular Arc

The **Arc Tools** are used to create arc curves by specifying points in the viewport. In Maya, there are two types of arc tools: **Three Point Circular Arc** and **Two Point Circular Arc**. To create an arc, choose **Create > Objects > Curve Tools** from the menubar; a cascading menu will be displayed. Choose **Two Point Circular Arc** from the cascading menu to create an arc by defining the start and end points of the arc. Similarly, choose the **Three Point Circular Arc** from the cascading menu to create an arc by defining the start point, the curve point, and the end point.

NURBS Curves and Surfaces

TOOLS FOR CREATING SURFACES

Maya provides a number of tools to create complex three dimensional surface models. To view the tools that are used to create various surfaces, select the **Modeling** option from the **Menuset** drop-down list in the Status Line. Next, choose the **Surfaces** menu to display all the surfacing tools in Maya, refer to Figure 3-16.

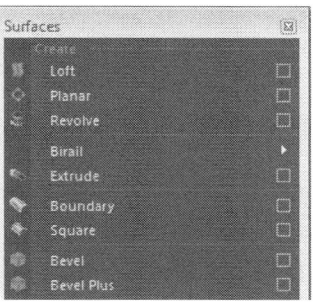

Figure 3-16 The **Surface** floating menu

Loft Tool

Menubar: Surfaces > Create > Loft

The **Loft** tool is used to skin a surface along the profile curves. While using this tool, at least two profile curves are required to create a NURBS surface. To create a NURBS surface by using this tool, create three curves, as shown in Figure 3-17. Next, press and hold the SHIFT key and select the curves in the viewport in order. Now, choose **Surfaces > Create > Loft** from the menubar; the NURBS curves are lofted with a surface in the viewport, as shown in Figure 3-18. To set the properties of the lofted surface created, choose **Surfaces > Create > Loft > Option Box** from the menubar; the **Loft Options** window will be displayed, as shown in Figure 3-19. The options in the **Loft Options** window are discussed next.

Figure 3-17 The NURBS curves before applying the **Loft** tool

Figure 3-18 The lofted surface created after applying the **Loft** tool

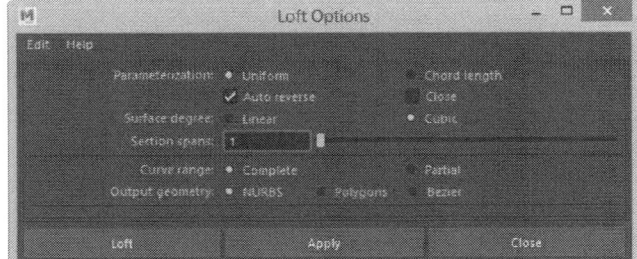

Figure 3-19 The **Loft Options** window

Planar Tool

Menubar: Surfaces > Create > Planar

The **Planar** tool is used to create a NURBS surface with all the vertices lying on the same plane. To create a NURBS surface using this tool, create a close curve using a curve tool. The curve should form a close loop and should at least have three sides. Next, choose **Surfaces > Create > Planar** from the menubar; a NURBS surface will be created. To set the properties of the NURBS surface, choose **Surfaces > Create > Planar > Option Box** from the menubar; the **Planar Trim Surface Options** window will be displayed, as shown in Figure 3-20. The options in this window are discussed next.

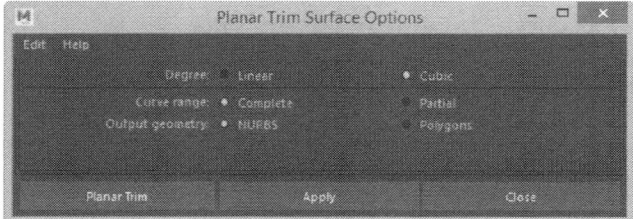

*Figure 3-20 The **Planar Trim Surface Options** window*

Revolve Tool

Menubar: Surfaces > Create > Revolve

The **Revolve** tool is used to create a surface around a profile curve along a selected axis. The axis of revolution depends on the location of the pivot point of an object. To create a revolved surface, choose **Create > Objects > Curve Tools > EP Curve Tool** from the menubar and then create a profile curve in the front-Z viewport, refer to Figure 3-21. Select the profile curve and choose **Surfaces > Create > Revolve** from the menubar; the profile curve will rotate around its pivot point, thus creating a revolved surface, as shown in Figure 3-22. Alternatively, you can choose **Surfaces > Create > Revolve > Option Box** from the menubar; the **Revolve Options** window will be displayed, as shown in Figure 3-23. The options in this window are discussed next.

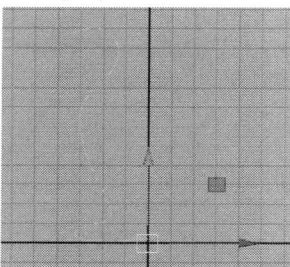

Figure 3-21 The profile curve created

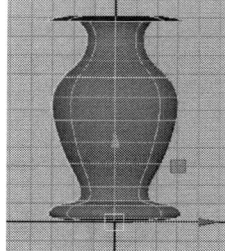

*Figure 3-22 The NURBS surface created after using the **Revolve** tool*

NURBS Curves and Surfaces

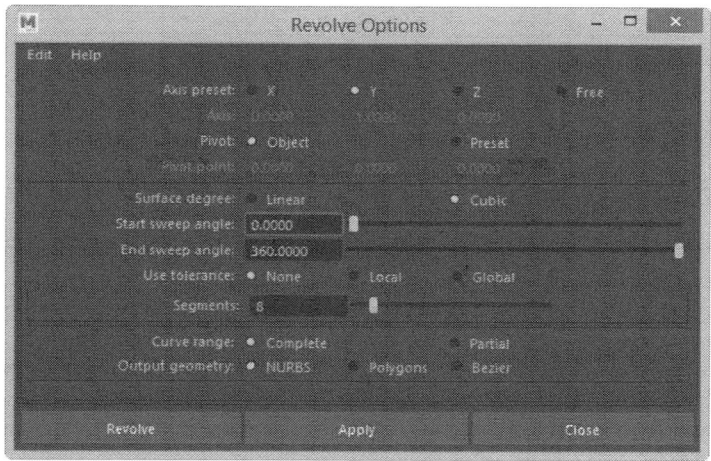

*Figure 3-23 The **Revolve Options** window*

Birail Tool

Menubar: Surfaces > Create > Birail

The **Birail** tool works similar to the **Extrude** tool. This tool is used to create surfaces using one curve or two profile curves along two path curves. You can create complex NURBS surfaces using this tool. Maya has three different types of **Birail** tools: **Birail 1 Tool**, **Birail 2 Tool**, and **Birail 3+ Tool**. Before creating a NURBS surface using different **Birail** tools, the following points should be kept in mind:

1. The profile curves and the path curves must touch each other and have continuity.

2. All profile curves should have the same number of CVs.

3. All path curves should also have the same number of CVs.

4. Press C to snap the curve of the profile curve and the path curve together.

5. If the profile curve and the path curve do not have the same number of CVs, you will have to draw the curves again.

Extrude Tool

Menubar: Surfaces > Create > Extrude

The **Extrude** tool is used to extrude a particular object by sweeping its profile curve along the path curve. To extrude a surface, two curves are required: a profile curve and a path curve. The profile curve gives shape to a surface, whereas the path curve defines the path on which the shape will sweep to create a surface. To create an extruded surface, select the two curves in the viewport. The first curve selected will act as the profile curve, whereas the second curve will act as the path curve. Now, choose **Surfaces > Create > Extrude** from the menubar to extrude the surface. You can use this method to create objects such as curtains, parts of a vehicle, and so on. To adjust the properties of the **Extrude** tool, choose **Surfaces > Create > Extrude > Option Box** from the menubar; the **Extrude Options** window will be displayed.

Boundary Tool

Menubar: Surfaces > Create > Boundary

The **Boundary** tool is used to create a surface by filling the boundary curves. This tool creates a NURBS surface by filling the space between the curves. It is not necessary for the curves to have a closed loop, but they should intersect with each other at some point. To apply the **Boundary** tool, create four curves in the viewport, as shown in Figure 3-24. Press and hold the SHIFT key and select all the curves in opposite pairs to maintain continuity. Now, choose **Surfaces > Create > Boundary** from the menubar to create the NURBS surface. To adjust the properties of the **Boundary** tool, choose **Surfaces > Create > Boundary > Option Box** from the menubar; the **Boundary Options** window will be displayed.

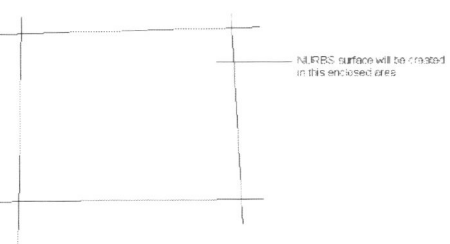

Figure 3-24 Four NURBS curves created

Square Tool

Menubar: Surfaces > Create > Square

The **Square** tool is used to create a four-sided NURBS surface from the intersecting curves. On choosing this tool, a NURBS surface is created by filling the region defined by four intersecting curves. This tool is similar to the **Boundary** tool with the only difference that in the **Boundary** tool, you can select curves in any order, whereas in the **Square** tool, you need to select them in clockwise or counterclockwise direction. To use this tool, create four intersecting curves in the viewport. Next, press and hold the SHIFT key and select the curves either in clockwise or counterclockwise direction. Now, choose **Surfaces > Create > Square** from the menubar; the NURBS surface will be created.

Bevel Tool

Menubar: Surfaces > Create > Bevel

The **Bevel** tool is used to create a NURBS surface by using the three-dimensional edge effect applied on the selected curves. The surface created by the **Bevel** tool has an open area that can be filled by using the **Planar** tool. To create a surface by using the **Bevel** tool, create a NURBS circle in the top-Y viewport, as shown in Figure 3-25. Next, choose **Surfaces > Create > Bevel** from the menubar; a beveled surface will be created, as shown in Figure 3-26. You can adjust the properties of the beveled surface in the **Channel Box / Layer Editor** by changing the values in the **bevel1** node of the **INPUTS** area as required, refer to Figure 3-27.

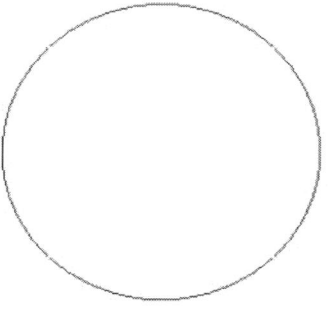

Figure 3-25 A NURBS circle

NURBS Curves and Surfaces

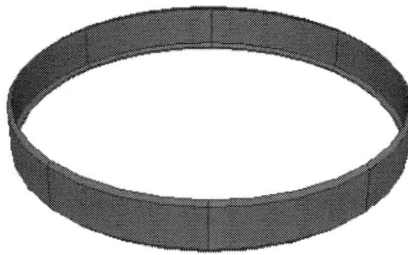

Figure 3-26 The bevel surface created

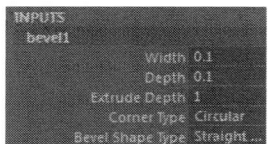

Figure 3-27 The **bevel1** node in the **INPUTS** area

Bevel Plus Tool

Menubar: Surfaces > Create > Bevel Plus

The **Bevel Plus** tool is used to extrude the closed curves and add beveled transition to the extruded surface. To create a surface by using this tool, create a NURBS circle in the top-Y viewport, as shown in Figure 3-28 and then choose **Surfaces > Create > Bevel Plus** from the menubar; a beveled surface will be created, as shown in Figure 3-29. You can adjust the properties of the beveled surface in the **Channel Box / Layer Editor** by changing the values in the **bevelPlus1** node of the **INPUTS** area as required, refer to Figure 3-30.

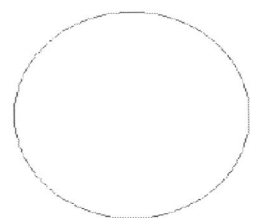

Figure 3-28 A NURBS circle created in the top-Y viewport

Figure 3-29 The beveled surface created using the **Bevel Plus** tool

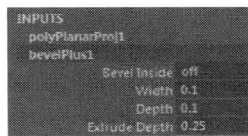

Figure 3-30 The **bevelPlus1** node displayed in the **INPUTS** area

TUTORIAL

All the files used in the tutorials can be downloaded from the CADSoft website (*www.cadsofttech.com*). These files are compressed in zip file format and are required to be extracted before using them in the tutorials. The path of the files is as follows: *Textbooks > Animation and Visual Effects > Maya > Autodesk Maya 2019 for Novices*

Tutorial 1

In this tutorial, you will create model of a 3D flower, as shown in Figure 3-31, using curve tools and the loft method.   **(Expected time: 30 min)**

Figure 3-31 The flower model

The following steps are required to complete this tutorial:

a. Create a project folder.
b. Create a profile shape.
c. Create leafs.
d. Change the background color of the scene.
e. Save and render the scene.

Creating a Project Folder

Create a new project folder with the name *c03_tut2* at *\Documents\maya2019* and then save the file with the name *c03tut2*, as discussed in Tutorial 1 of Chapter 2.

Creating a Profile Shape

In this section, you will create a profile shape of the flower using the **Circle** tool.

1. Turn off the **Interactive Creation** option as discussed earlier. Choose the **Four View** button from the Tool Box to switch to four views. Move the cursor to the top-Y viewport and then press the SPACEBAR key to maximize the top-Y viewport. Choose **Create > NURBS Primitives > Circle > Option Box** from the menubar; the **NURBS Circle Options** window is displayed in the viewport. Enter required values in the **NURBS Circle Options** window, as shown in Figure 3-32.

2. Choose **Edit > Duplicate > Duplicate** from the menubar; a copy of the circle is created in the viewport. Select **Move Tool** from the Tool Box and then move the duplicate circle along the Y-axis in the persp viewport.

NURBS Curves and Surfaces

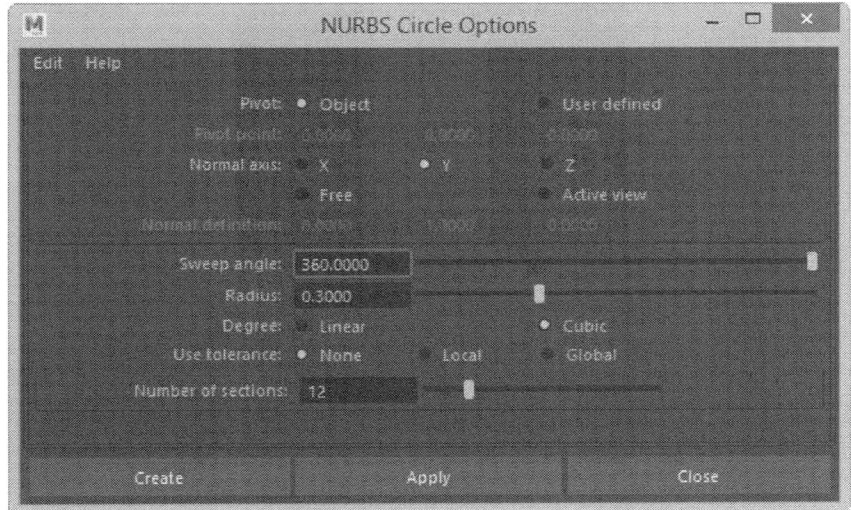

*Figure 3-32 The **NURBS Circle Options** window*

3. Press and hold the right mouse button and choose **Control Vertex** from the marking menu. The control vertices will be displayed in the viewport. Select every second vertex in the top-Y viewport, as shown in Figure 3-33.

4. Choose **Scale Tool** from the Tool Box and scale the selected control vertices, as shown in Figure 3-34.

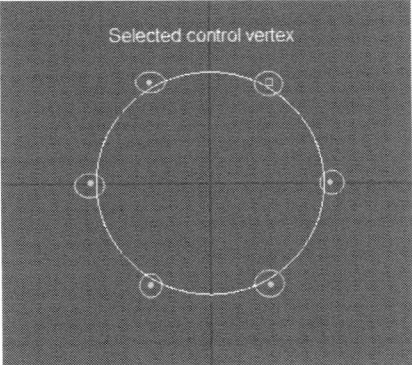

Figure 3-33 Selected control vertices

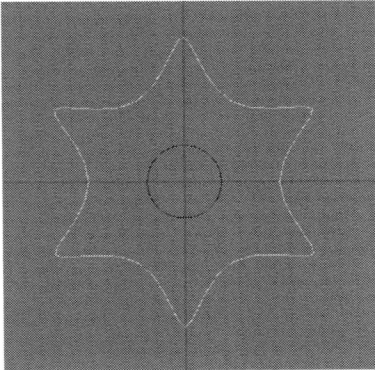

Figure 3-34 Control vertices after scaling

5. Create 5 copies of the modified circle and then align them., as shown in Figures 3-35 and 36.

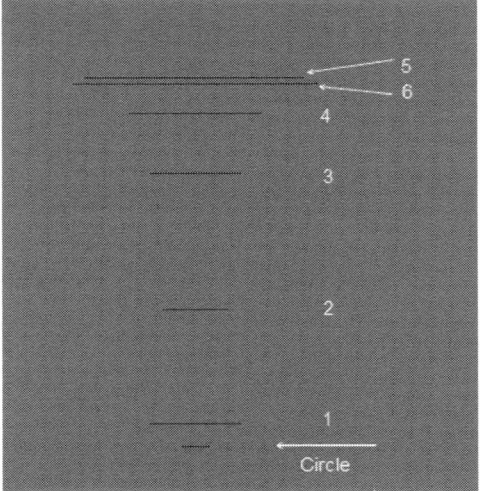 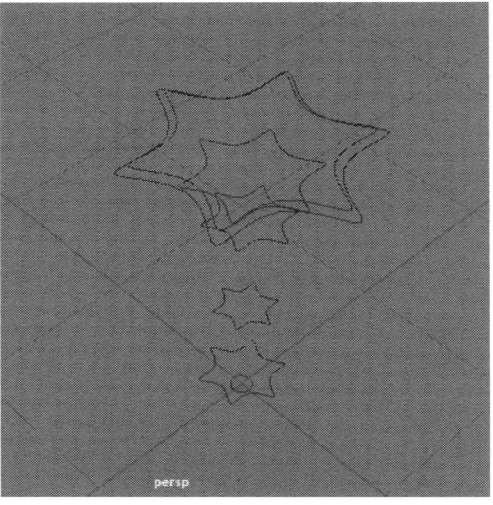

Figure 3-35 The aligned shapes in the viewport

Figure 3-36 Shapes after duplicate in persp viewport

6. First select the shape marked as 1 in Figure 3-61 and then select other shapes using the SHIFT key in the order. Choose **Surfaces > Create > Loft** from the menubar; a surface is created, as shown in Figure 3-37.

 Note
*By default, two side lighting option is not enabled in Maya. As a result, the inner surface appears black in the viewport. To view the objects in uniform shading, choose **Lighting > Two Sided Lighting** from the **Panel** menu.*

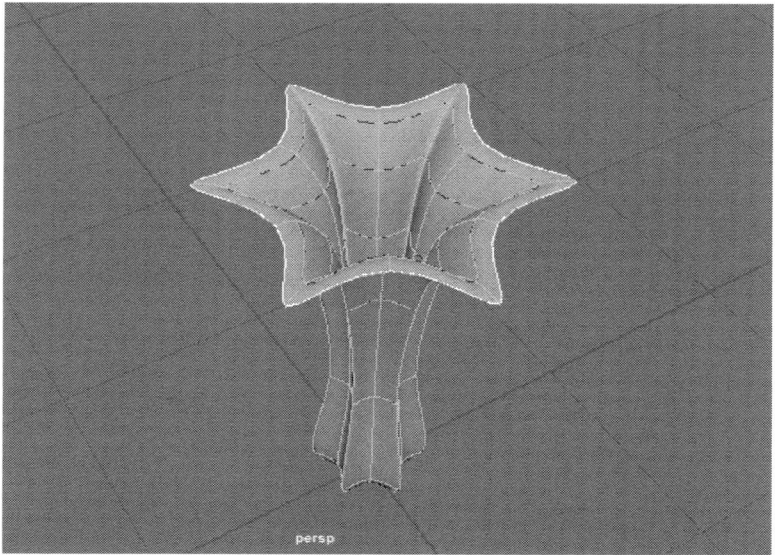

Figure 3-37 Surface created

NURBS Curves and Surfaces

7. Now, you can select the curves and scale them as per requirement, refer to Figure 3-38.

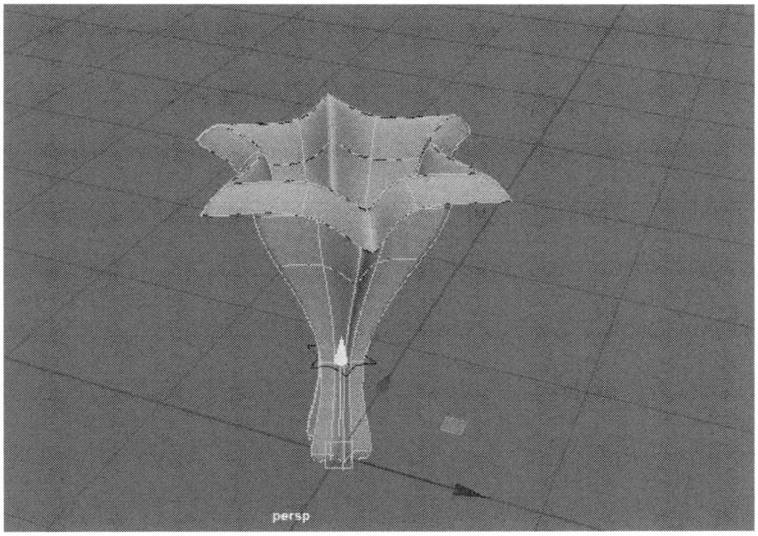

Figure 3-38 The scaled shape of the flower

Creating Leaves

In this section, you will create leaves of the flower using **CV Curve /Tool**.

1. Maximize the top-Y viewport. Choose **Create > Curve Tools > CV Curve Tool > Option Box** from the menubar; the **Tool Settings (CV Curve Tool)** panel is displayed. Select the **5** radio button corresponding to the **Curve degree** attribute.

2. Create **3** profile curves for the leaf in the top-Y viewport, as shown in Figure 3-39.

3. Activate the persp viewport, select profile 1 and right click on it; a marking menu is displayed. Choose **Control Vertex** from the marking menu. Now, press SHIFT and select profile 2 and choose **Control Vertex** from the marking menu. Repeat the process for third profile as well.

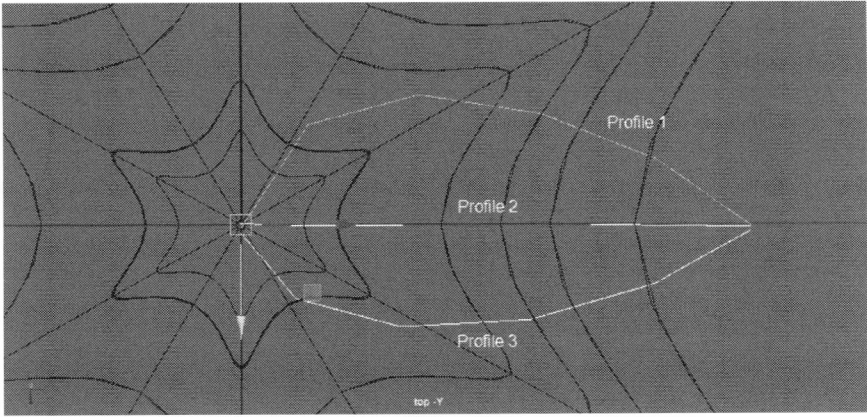

Figure 3-39 Profile curves for leaf

4. Modify the shapes of leaf using control vertices, as shown in Figure 3-40.

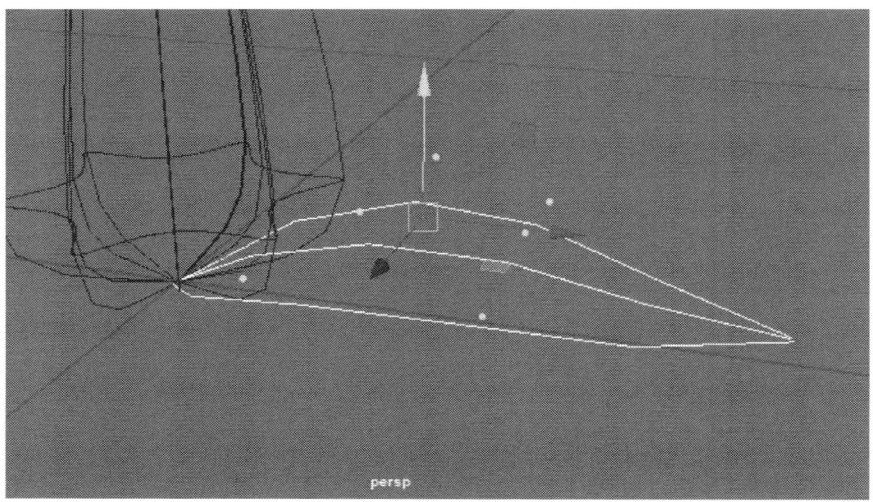

Figure 3-40 Profile curves modified

5. Now, select profiles in an order using SHIFT and then choose **Surfaces > Create > Loft** from the menubar; a surface is created the selected profile curves, refer to Figure 3-41. If leaf appears smaller in size, select surface and scale it by using **Scale Tool** from the Tool Box.

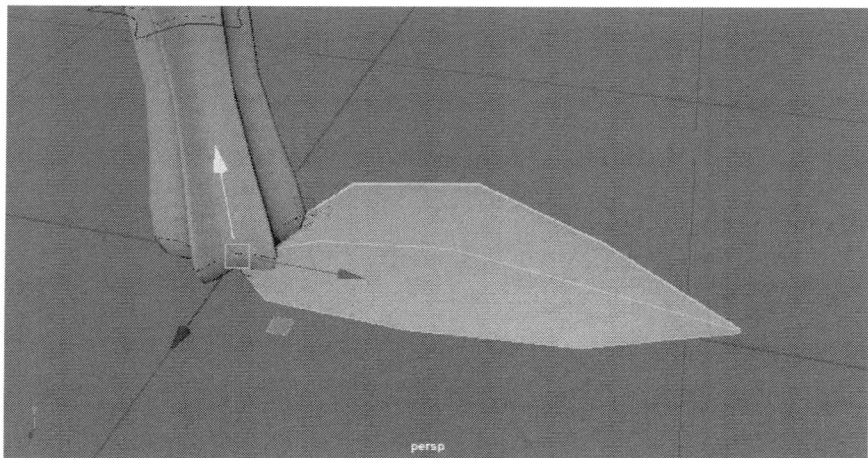

Figure 3-41 Leaf created by using loft

6. Select leaf in viewport, choose **Edit > Duplicate > Duplicate** from the menubar. Choose **Rotate Tool** from the Tool Box and rotate leaf along the Y axis. Similarly, duplicate again and rotate and align it, refer to Figure 3-42.

NURBS Curves and Surfaces

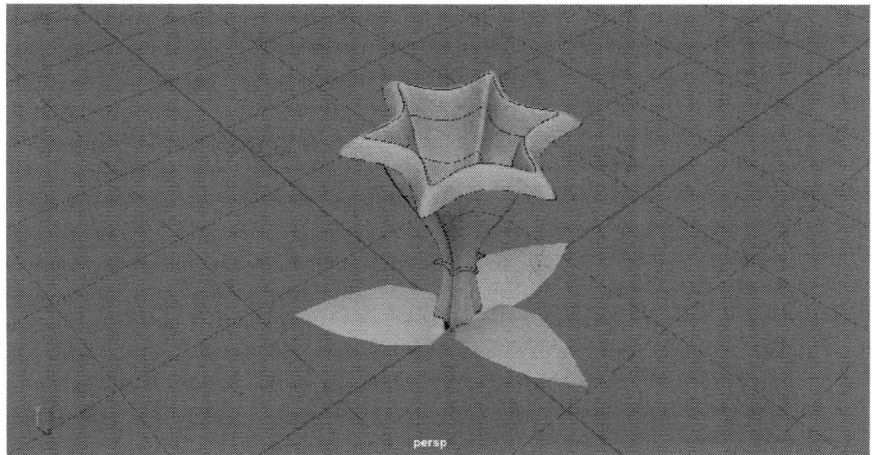

Figure 3-42 Leaves created by using loft

Changing the Background Color of the Scene
In this section, you will change the background color of the scene.

1. Choose **Windows > Editors > Outliner** from the menubar; the **Outliner** window is displayed. Select the **persp** camera in the **Outliner** window; the **perspShape** tab is displayed in the **Attribute Editor**.

2. In the **perspShape** tab, expand the **Environment** area and drag the **Background Color** slider bar toward right to change the background color to white.

Saving and Rendering the Scene
In this section, you will save the scene that you have created and then render it. You can view the final rendered image of the scene by downloading the *c03_maya_2019_rndr.zip* file from *www.cadsofttech.com*. The path of the file is mentioned in Tutorial 1.

1. Choose **File > Save Scene** from the menubar.

2. Maximize the persp viewport, if it is not already maximized. Choose the **Render the current frame** button from the Status Line; the **Render View** window is displayed. In this window, change renderer to **Maya Software** and choose the **Redo previous render** button available below the **File** menu. This window shows the final output of the scene.

EXERCISES
The rendered output of the models used in the following exercises can be accessed by downloading the *c03_maya_2019_exr.zip* file from *www.cadsofttech.com*. The path of the file is as follows: *Textbooks > Animation and Visual Effects > Maya > Autodesk Maya 2019 for Novices*.

Exercise 1

Create the model of an apple, as shown in Figure 3-43.   **(Expected time: 15 min)**

Figure 3-43 Model of an apple

Exercise 2

Create the model of a lantern, as shown in Figure 3-44.   **(Expected time: 15 min)**

Figure 3-44 Model of a lantern

Chapter 4

NURBS Modeling

Learning Objectives

After completing this chapter, you will be able to:
- *Understand NURBS editing techniques*
- *Convert NURBS objects to polygons*

INTRODUCTION

NURBS stands for Non Uniform Rational B-Spline. NURBS are used for creating 3D curves and surfaces, and complex 3D organic models having smooth surfaces and curves. In the previous chapter, you have learned about different methods of creating NURBS surfaces. In this chapter, you will learn about various editing techniques used for modifying NURBS surfaces.

WORKING WITH NURBS TOOLS

The NURBS tools are used to edit NURBS surfaces. The most commonly used tools in NURBS modeling are discussed next.

Duplicate NURBS Patch

| Menubar: | Surfaces > Edit NURBS Surfaces > Duplicate NURBS Patch |

The **Duplicate NURBS Patch** tool is used to create new surface from an existing NURBS patch. To understand the working of this tool, choose **Create > Objects > NURBS Primitives > Sphere** from the menubar and create a NURBS sphere in the viewport. Press and hold the right mouse button over the sphere and then choose **Surface Patch** from the marking menu displayed, as shown in Figure 4-1; the surface patch component of the NURBS sphere will be activated. Now, select the surface patch that you want to duplicate. Choose **Surface > Edit NURBS Surfaces > Duplicate NURBS Patch** from the menubar; a duplicate surface patch will be created. Invoke **Move Tool** from the Tool Box and move the duplicate surface patch away from the NURBS sphere. Note that the pivot point of the duplicate surface patch will remain at the same position as that of the NURBS sphere. To reset the pivot point to the center of the duplicate patch, choose **Modify > Pivot > Center Pivot** from the menubar; the pivot point will be reset.

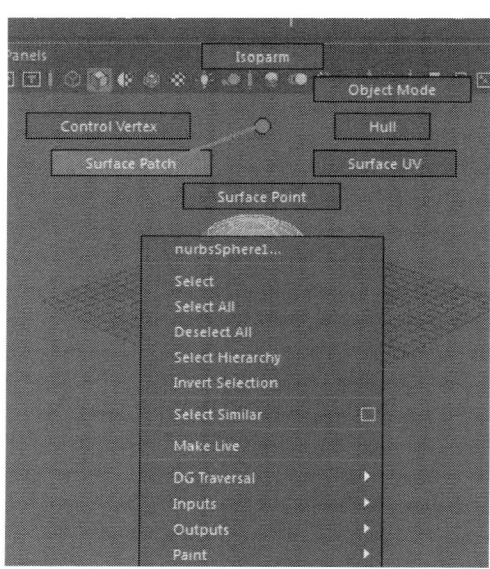

*Figure 4-1 Choosing **Surface Patch** from the marking menu*

Project Curve on Surface

| Menubar: | Surfaces > Edit NURBS Surfaces > Project Curve on Surface |

The **Project Curve on Surface** tool is used to project a NURBS curve on a NURBS surface. To understand the working of this tool, choose **Create > Objects > NURBS Primitives > Square** from the menubar and create a square in the front-Z viewport. Now, choose **Surfaces > Create > Planar** from the menubar to create a NURBS surface. Next, choose **Create > Curve Tools > EP Curve Tool** from the menubar to create a curve, as shown in Figure 4-2 and make sure the NURBS curve is selected. Press and hold the SHIFT key and select the NURBS surface. Now, choose **Surfaces > Edit NURBS Surfaces > Project Curve on Surface** from the menubar to project the curve on the surface and activate the persp viewport; the NURBS curve will be projected over the NURBS surface, as shown in Figure 4-3.

NURBS Modeling

Note
The curve will be projected at the exact position as visible through the camera of that particular viewport.

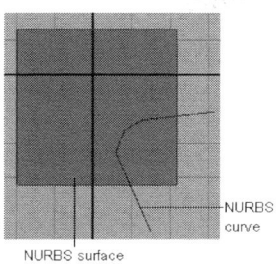

Figure 4-2 The NURBS curve created

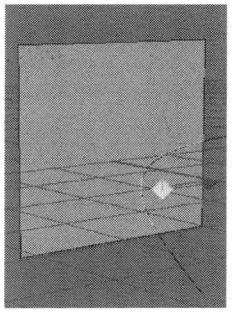

Figure 4-3 The NURBS curve projected on the NURBS surface

Intersect

Menubar: Surfaces > Edit NURBS Surfaces > Intersect

The **Intersect** tool is used to create a new segment at the intersection of two NURBS surfaces. To understand the working of this tool, consider two surfaces, as discussed earlier, and align them such that they intersect each other, refer to Figure 4-4. Select both the surfaces and choose **Surfaces > Edit NURBS Surfaces > Intersect** from the menubar to create a new segment at the location where the two plane intersect. Choose the **Move Tool** from the Tool Box; the selection handles will be displayed at the intersection point, refer to Figure 4-4. You can now move these handles to align the intersection point anywhere on the NURBS surface.

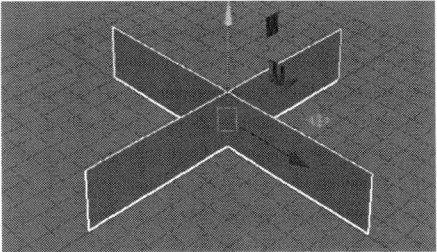

Figure 4-4 The aligned surfaces and selection handles

Trim Tool

Menubar: Surfaces > Edit NURBS Surfaces > Trim Tool

The **Trim Tool** is used to hide a particular area from the NURBS surface defined by curves. To understand the working of this tool, consider a curve projected on a surface as discussed in the **Project Curve on Surface** section. Next, select the NURBS surface in the viewport and choose **Surfaces > Edit NURBS Surfaces > Trim Tool** from the menubar; the NURBS surface will be displayed in the wireframe mode with a dotted outline. Select the part that you want to retain from the surface and press ENTER; the surface will be trimmed. You can also change the settings of **Trim Tool** as required. To retain the selected part from the NURBS surface and trim the unselected part from the NURBS surface, choose **Surfaces > Edit NURBS Surfaces > Trim Tool > Option Box** from the menubar; the **Tool Settings (Trim Tool)** panel will be displayed on the viewport, as shown in Figure 4-5. The **Keep** radio button is used. The **Discard** radio button is used to trim the selected part from the NURBS surface and keep the unselected part intact.

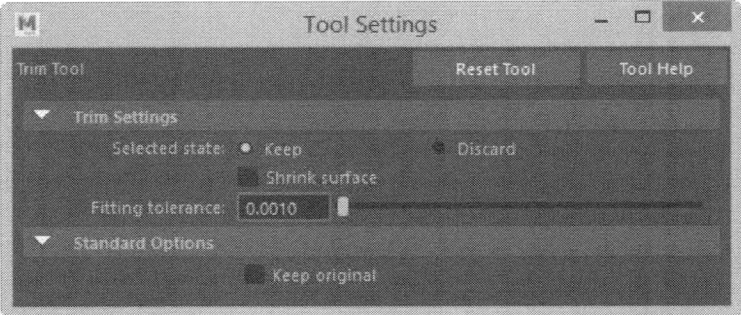

*Figure 4-5 The **Tool Settings (Trim Tool)** panel*

Untrim

Menubar: Surfaces > Edit NURBS Surfaces > Untrim

The **Untrim** tool is used to untrim the last trimmed surface. To understand the working of this tool, select the trimmed surface and choose **Surfaces > Edit NURBS Surfaces > Untrim** from the menubar; the surface sets back to its original untrimmed state.

Attach

Menubar: Surfaces > Edit NURBS Surfaces > Attach

The **Attach** tool is used to attach two selected NURBS surfaces. To understand the working of this tool, create two NURBS surfaces in the viewport, as shown in Figure 4-6. Next, select the two surfaces and choose **Surface > Edit NURBS Surfaces > Attach > Option Box** from the menubar; the **Attach Surfaces Options** window will be displayed, as shown in Figure 4-7. By default, the **Blend** radio button is selected in the **Attach method** area. Choose the **Apply** button; the selected surfaces will be connected, as shown in Figure 4-8. You can also select the **Connect** radio button from the **Attach** method area to connect the end of a surface to the end of another surface, as shown in Figure 4-9.

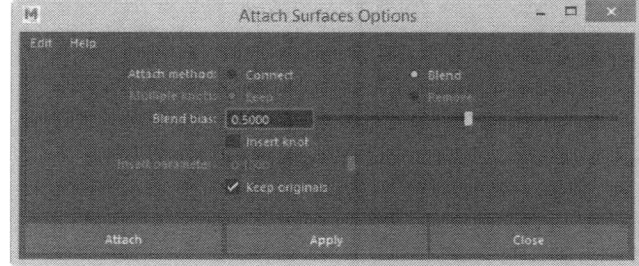

Figure 4-6 The two NURBS surfaces   *Figure 4-7 The **Attach Surfaces Options** window*

Note
*The **Attach Surfaces** tool does not attach trimmed surfaces. In such cases, surfaces need to be untrimmed before attaching them using this tool.*

NURBS Modeling

Figure 4-8 Surfaces connected on selecting the **Blend** radio button

Figure 4-9 The surfaces connected on selecting the **Connect** radio button

Tip
You can make changes in the attached surfaces by using the **attachSurface1** tab in the **Attribute Editor**. To do so, choose **Window > Editors > General Editors > Attribute Editor** from the menubar; the **Attribute Editor** will be displayed. Choose the **attachSurface#** tab from the **Attribute Editor**; the attributes of the attached surface will be displayed, as shown in Figure 4-10. You can apply different styles on the surface by using the parameters in the **Attribute Editor**.

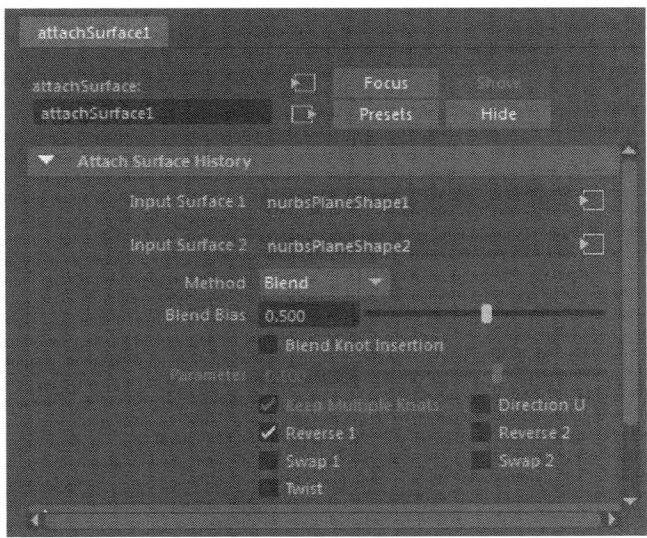

Figure 4-10 The *attachSurface1* tab in the **Attribute Editor**

Attach Without Moving

Menubar: Surfaces > Edit NURBS Surfaces > Attach Without Moving

The **Attach Without Moving** tool is used to attach two NURBS surfaces or curves by selecting their respective isoparms. To understand the working of this tool, create two NURBS surfaces and select two isoparms, one each from the two surfaces. Now, choose **Surfaces > Edit NURBS**

Surfaces > Attach Without Moving from the menubar; the two surfaces will be attached along with the two isoparms.

Align

Menubar: Surfaces > Edit NURBS Surfaces > Align

The **Align** tool is used to align the selected NURBS surfaces tangentially. To understand the working of this tool, select the NURBS surfaces that you want to align. Choose **Surfaces > Edit NURBS Surfaces > Align > Option Box** from the menubar; the **Align Surfaces Options** window will be displayed, as shown in Figure 4-11. Now, you can use different options in this window to align the selected NURBS surfaces as required.

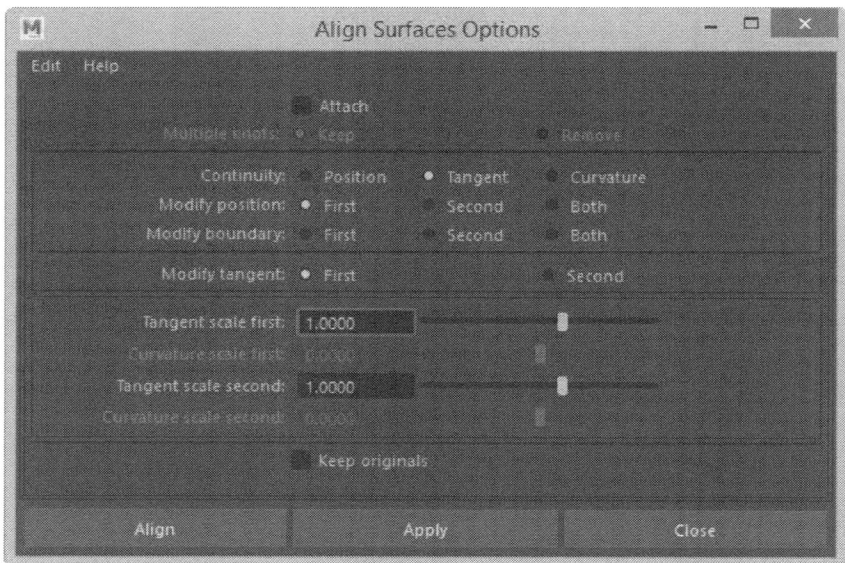

Figure 4-11 The Align Surfaces Options window

Detach

Menubar: Surfaces > Edit NURBS Surfaces > Detach

The **Detach** tool is used to break NURBS surfaces into parts. To understand the working of this tool, select the isoparm of a surface and then choose **Surfaces > Edit NURBS Surfaces > Detach** from the menubar; the surface will get detached from the selected isoparm.

Open/Close

Menubar: Surfaces > Edit NURBS Surfaces > Open/Close

The **Open/Close** tool is used to open or close the NURBS surfaces. To open a closed surface, select the closed surface and choose **Surfaces > Edit NURBS Surfaces > Open/Close** from the menubar; the closed surface will be opened, as shown in Figure 4-12. Similarly, to close an open surface, select the opened surface in the viewport and choose **Surfaces > Edit NURBS**

Surfaces > Open/Close from the menubar; the opened surface will change into a closed surface, as shown in Figure 4-13.

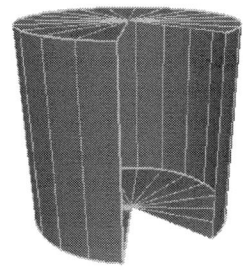

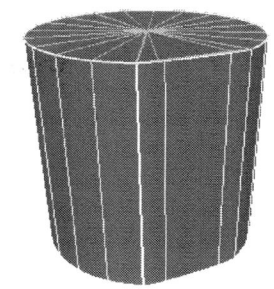

Figure 4-12 The opened NURBS surface   *Figure 4-13* The closed NURBS surface

Push

The **Push** option is used to push down the selected NURBS mesh. To do so, consider a NURBS plane in the viewport. Next, choose **Surfaces > Edit NURBS Surfaces > Sculpt Geometry Tool > Option Box** from the menubar; the **Tool Settings (Sculpt Geometry Tool)** panel will be displayed on the viewport. Choose the **Push** button in the **Sculpt Parameters** area; the cursor will be displayed, as shown in Figure 4-14. Press and hold the left mouse button and move the cursor over the NURBS plane to sculpt the NURBS plane.

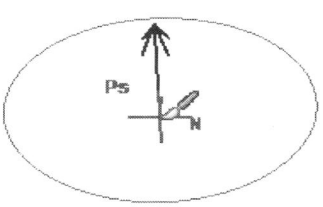

Figure 4-14 The **Sculpt Geometry Tool** cursor

Pinch

The **Pinch** tool is used to pull selected vertices toward each other. It helps in bringing the vertices closer in order to make sharp or well defined creases.

Slide

The **Slide** tool is used to slide the vertices of the surface in the direction of the stroke Figure 4-15 shows a surface before sliding the vertices and Figure 4-16 shows the surface after sliding the vertices.

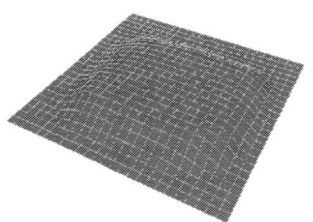

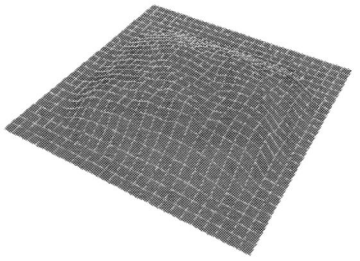

Figure 4-15 Surface before sliding   *Figure 4-16* Surface after sliding

Erase

The **Erase** tool is used to erase the changes made on the surface by using the other **Sculpt Geometry** tools like **Push** or **Pull**. Figure 4-17 shows a surface before erasing changes on it and Figure 4-31 shows the surface after erasing changes on it.

Figure 4-17 Surface after erasing

CONVERTING OBJECTS

In Maya, you can convert the form of an object. To do so, select an object in the viewport and choose **Modify > Objects > Convert** from the menubar; a cascading menu will be displayed. Choose the conversion type from the cascading menu to specify the output geometry for the selected object. The most commonly used options in this cascading menu are discussed next.

Converting NURBS to Polygons

Menubar:	Modify > Objects > Convert > NURBS to Polygons

The **NURBS to Polygons** conversion tool is used to convert a NURBS mesh into a polygonal object. To do so, select a NURBS mesh to be converted and then choose **Modify > Objects Convert > NURBS to Polygons > Option Box** from the menubar; the **Convert NURBS to Polygons Options** window will be displayed in the viewport, as shown in Figure 4-18. You can use this window to set the options for the conversion of the object from NURBS to polygons.

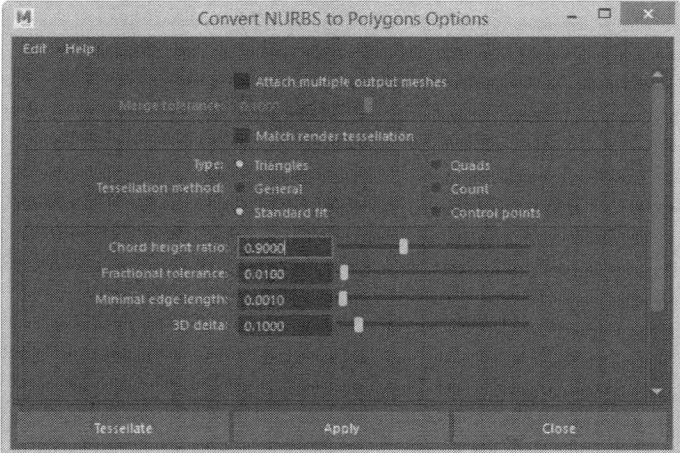

*Figure 4-18 The **Convert NURBS to Polygons Options** window*

NURBS Modeling

TUTORIAL

All the files used in the tutorials can be downloaded from the CADSoft website (*www.cadsofttech.com*). These files are compressed in zip file format and are required to be extracted before using them in the tutorials. The path of the files is as follows: *Textbooks > Animation and Visual Effects > Maya > Autodesk Maya 2019 for Novices*

Tutorial 1

In this tutorial, you will create the model of a cowboy hat, as shown in Figure 4-19, using NURBS. **(Expected time: 15 min)**

Figure 4-19 The model of a cowboy hat

The following steps are required to complete this tutorial:

a. Create a project folder.
b. Create a NURBS cylinder.
c. Add details to the hat.
d. Change background color of the scene.
e. Save and render the scene.

Creating a Project Folder

Create a new project folder with the name *c04_tut1* at *\Documents\maya2019* and then save the file with the name *c04tut1*, as discussed in Tutorial 1 of Chapter 2.

Creating a NURBS Cylinder

In this section, you will create a NURBS cylinder to form the base structure of the cowboy hat.

1. Maximize the top-Y viewport and then choose **Create > Objects > NURBS Primitives > Cylinder > Option Box** from the menubar; the **NURBS Cylinder Option** window is displayed on the viewport, set the values of the parameters as shown in Figure 4-20.

2. Select **nurbsCylinder1** on the viewport and rename **nurbsCylinder1** to *hat*.

3. In the front-Z viewport, press and hold the right mouse button on the *hat*; a marking menu is displayed. Choose **Control Vertex** from the marking menu; the vertex selection mode is activated. Now, select the vertices, as shown in Figure 4-21. Next, invoke **Scale Tool** from the Tool Box and scale up the vertices uniformly in the front-Z viewport, as shown in Figure 4-22.

4. Make sure the vertices of *hat* are selected. Select the green handle of **Scale Tool** and scale the selected vertices downward along the Y-axis; the mesh gets modified, as shown in Figure 4-23.

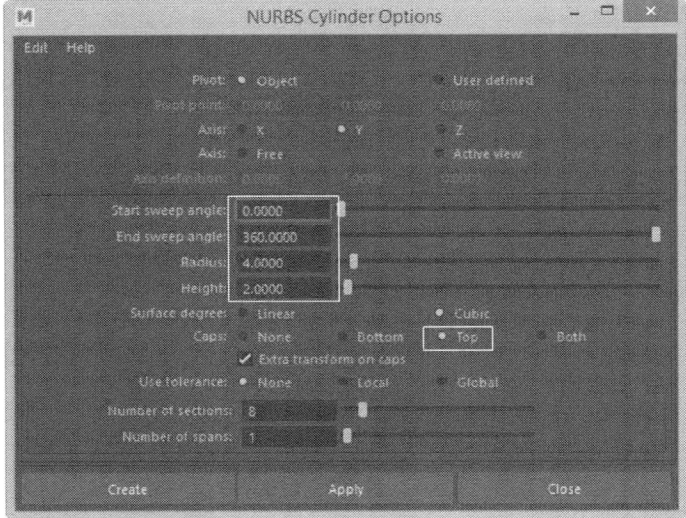

Figure 4-20 The **NURBS Cylinder Options** *window*

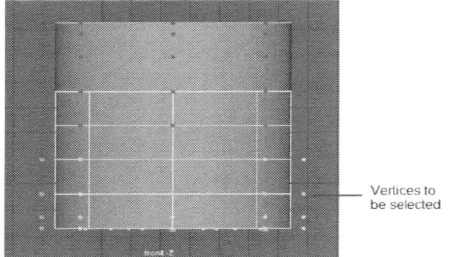

Figure 4-21 Vertices to be selected

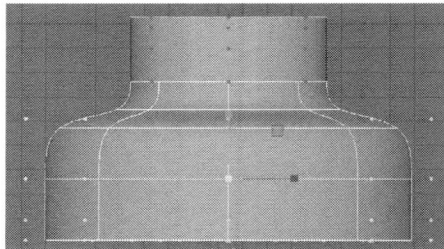

Figure 4-22 Scaling the selected vertices uniformly

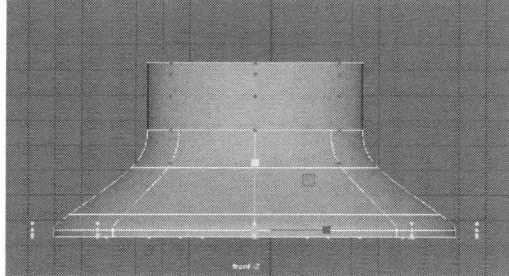

Figure 4-23 Scaling the selected vertices along the Y axis

Adding Details to the Hat

In this section, you will add details to the hat to give it the look of a cowboy hat.

1. Maximize the top-Y viewport. Next, press and hold the SHIFT key and marquee-select the vertices, refer to Figure 4-24. Choose **Scale Tool** from the Tool Box, and scale the selected

NURBS Modeling

vertices inward along the Z-axis using the yellow handle; the mesh gets scaled, as shown in Figure 4-25.

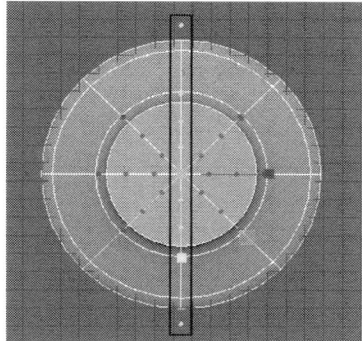
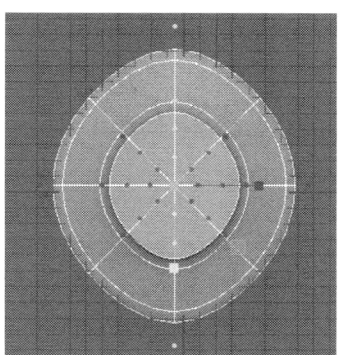

Figure 4-24 The vertices selected   *Figure 4-25* Scaled mesh

2. Maximize the side-X viewport. Next, marquee-select the vertices using the SHIFT key, as shown in Figure 4-26. Next, choose **Move Tool** and move the vertices downward along the Y-axis, as shown in Figure 4-27.

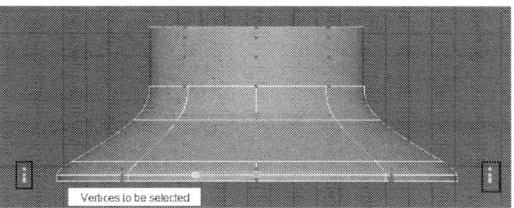
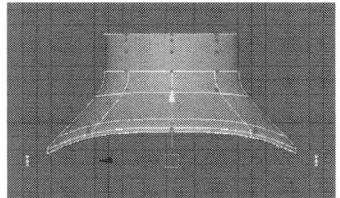

Figure 4-26 Vertices to be selected   *Figure 4-27* The modified mesh after moving the vertices

3. Maximize the persp viewport. Press and hold the right mouse button over the lower part of *hat*; a marking menu is displayed. Next, choose **Object Mode** from the marking menu; the object selection mode is activated.

4. Select the model in the persp viewport. Next, press and hold the right mouse button over the lower part of *hat* and choose **Isoparm** from the marking menu; the color of the edges of *hat* turns blue.

5. In the persp viewport, select the isoparm, as shown in Figure 4-28. Drag the isoparm outward; a dotted isoparm is displayed on *hat*. Next, choose the **Modeling** menuset from the **Menuset** drop-down list in the Status Line and then choose **Surfaces > Edit NURBS Surfaces > Insert Isoparms** from the menubar; a new isoparm is added, as shown in Figure 4-29.

6. Make sure *hat* is selected. Next, choose **Edit > Delete > Delete All by Type > History** from the menubar; the history of all actions performed on the model is deleted.

7. Maximize the side-X viewport. Next, press and hold the right mouse button on *hat*; a marking menu is displayed. Choose **Control Vertex** from the marking menu; the vertex selection mode is activated. Next, marquee-select the vertices of *hat* by using the SHIFT key, as shown

in Figure 4-30. Next, invoke **Move Tool** from the Tool Box, and move the vertices upward along the Y axis, as shown in Figure 4-31.

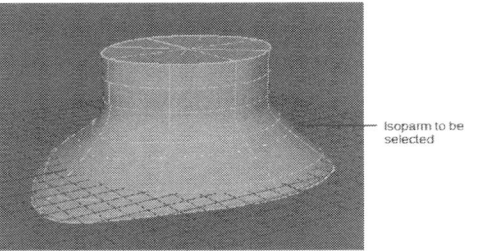
Figure 4-28 Selecting an isoparm

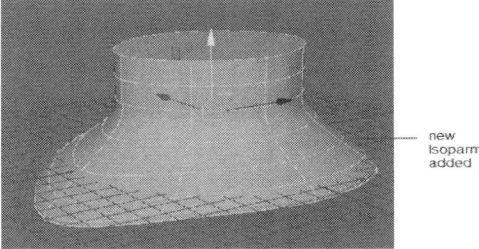
Figure 4-29 A new isoparm added

8. Press and hold the right mouse button over *hat* and choose **Object Mode** from the marking menu displayed; the object selection mode is activated. Maximize the top-Y viewport. Choose **View > Predefined Bookmarks > Bottom** from the **Panel** menu; the bottom viewport is activated.

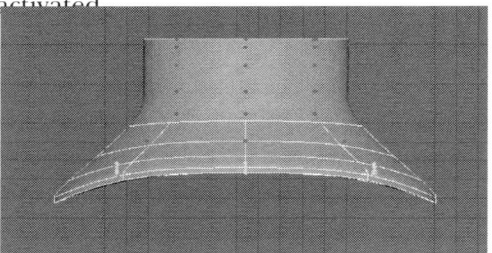
Figure 4-30 Vertices to be selected

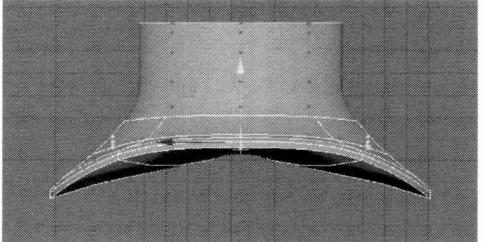
Figure 4-31 The modified mesh

9. Insert two new isoparms on *hat*, as discussed in steps 4 and 5. Figure 4-32 displays two isoparms added to *hat* in bottom viewport.

10. Press and hold the right mouse button over the cylinder and choose **Control Vertex** from the marking menu displayed; the vertex selection mode is activated. Next, select the vertices using the SHIFT key, as shown in Figure 4-33.

11. Choose **View > Predefined Bookmarks > Right Side** from the **Panel** menu; the right side-X viewport is activated. Choose **Move Tool** from the Tool Box and move the selected vertices upward along the Y axis to get the final output, refer to Figure 4-34.

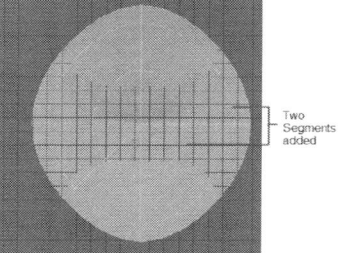
Figure 4-32 The two new isoparms added

Changing the Background Color of the Scene
In this section, you will change the background color of the scene.

1. Choose **Windows > Editors > Outliner** from the menubar; the **Outliner** window is displayed. Select the **persp** camera in the **Outliner** window; the **perspShape** tab is displayed in the **Attribute Editor**.

NURBS Modeling

2. In the **perspShape** tab, expand the **Environment** area and drag the **Background Color** slider bar toward right to change the background color to white.

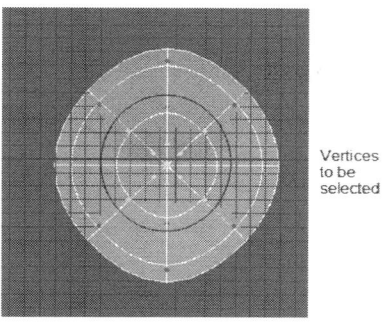

Figure 4-33 Selecting vertices

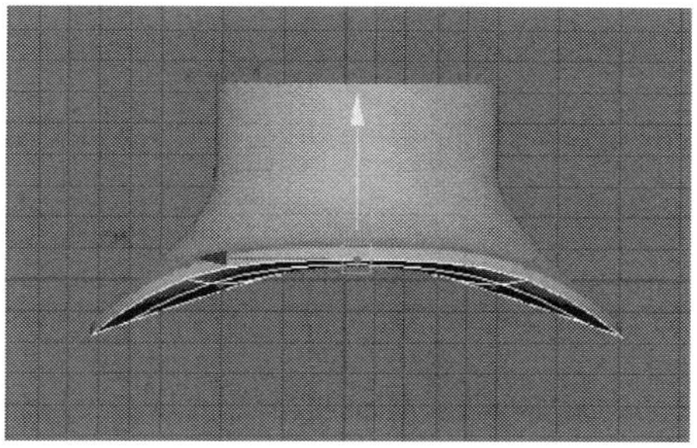

Figure 4-34 The final output

Saving and Rendering the Scene

In this section, you will save the scene that you have created and then render it. You can view the final rendered image of this scene by downloading the *c04_maya_2019_rndr.zip* file from *www.cadsofttech.com*. The path of the file is as follows: *Textbooks > Animation and Visual Effects > Maya > Autodesk Maya 2019 for Novices*.

1. Choose **File > Save Scene** from the menubar.

2. Maximize the persp viewport if not already maximized. Choose the **Render the current frame** button from the Status Line; the **Render View** window is displayed. This window shows the final output of the scene, refer to Figure 4-19.

EXERCISES

The rendered output of the models used in the following exercises can be accessed by downloading the *c04_maya_2019_exr.zip* from *www.cadsofttech.com*. The path of the file is as follows: *Textbooks > Animation and Visual Effects > Maya > Autodesk Maya 2019 for Novices*.

Exercise 1

Use various NURBS modeling techniques to create the model of a handbag, as shown in Figure 4-35. **(Expected time: 45 min)**

Figure 4-35 The model of a handbag

Exercise 2

Use various NURBS modeling techniques to create the model of a chair, as shown in Figure 4-36. **(Expected time: 30 min)**

Figure 4-36 The model of a chair

Chapter 5

UV Mapping

Learning Objectives

After completing this chapter, you will be able to:
- *Use different UV mapping techniques*
- *Use the UV Editor*
- *Use various tools and options in the UV Editor*

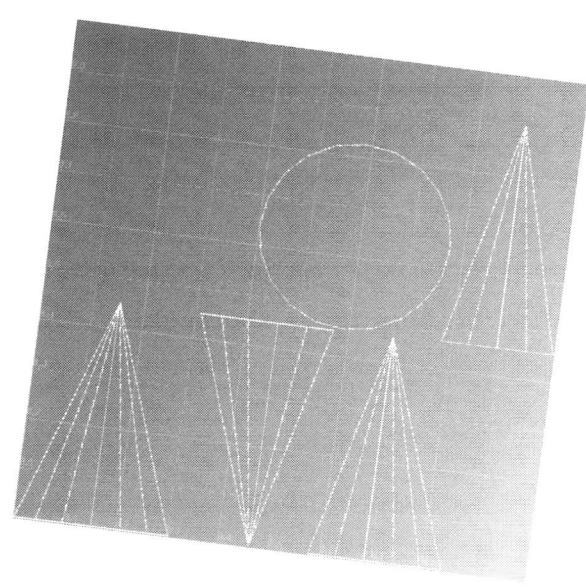

INTRODUCTION

UV mapping is a process of creating or editing UVs for an object, where U and V denote the axes of 2D texture and determine how the texture will be mapped to the surface of an object. In Maya, various types of UV mapping techniques are used to apply texture on an object. In this chapter, you will learn about the tools and techniques used in Maya to create and apply different UV maps.

UV MAPPING

UV mapping is a technique in which a 3D object is unfolded and split into 2D patches. It is used to place texture directly on the surface mesh. The UV coordinates are used to position textures on the surfaces. To access a UV mapping technique, select **Modeling** from the **Menuset** drop-down list in the Status Line. Next, choose the required mapping technique from the **UV** menu of the menubar. There are six types of UV mapping used in Maya and all of them are discussed in the chapter.

UV EDITOR

Menubar: UV > UV Editor

The **UV Editor**, as shown in Figure 5-1, is used to view and edit the UV texture coordinates within a 2D view.

Tip
To pan in the UV Editor, press and hold the middle mouse button along with the ALT key.

To view the UV coordinates of an object, create in the viewport and then select it. Next, choose **UV > UV Editor** from the menubar; **UV Editor** will be displayed with the UV texture coordinates of the object. Figure 5-1 shows the UV coordinates of a cube primitive in the **UV Editor**. In the **UV Editor**, the tools are grouped together in the toolbar and are discussed next.

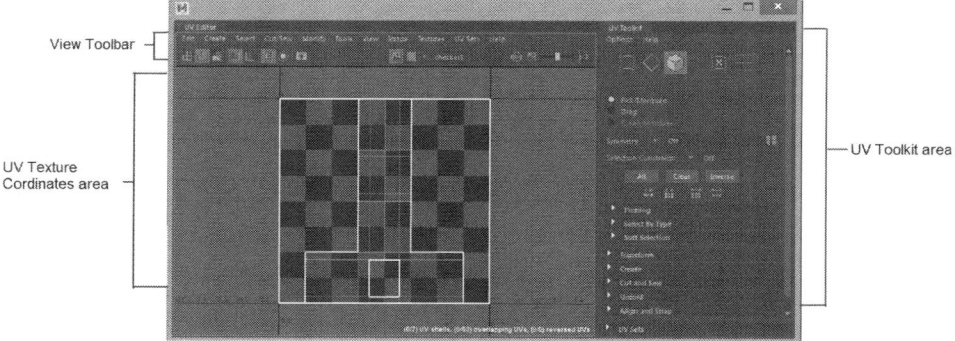

Figure 5-1 The UV Editor

UV Mapping

Note
*You can select UVs and UV shells in the scene as well as in the **UV Editor**. However, you can only edit UVs in the **UV Editor**. To select UVs in the scene, right-click on a mesh; a marking menu will displayed. To select UVs, choose **UV > UV** from the marking menu. To select UV shells, choose **UV > UV Shell** from the marking menu. Now, click on the UVs to select individual UVs or marquee drag to select a region of UVs. When UVs are selected in the scene, CTRL+right-click on the mesh and choose **Grow Selection** or **Shrink Selection** from the marking menu to grow or shrink the UV selection. To select UVs in the **UV Editor**, right-click on the 2D view of the **UV Editor** and then select a component mode from the marking menu. Now, click or marquee drag to select the component.*

View Toolbar
The tools in the View Toolbar of the **UV Editor** are used to change the display of the UV Shells in the **UV Editor** and the viewport. Some of the tools are discussed next.

Wireframe
| **UV Editor menubar:** | View > Wireframe |

This tool is used to display the UVs in wireframe. The background will be transparent. You can also choose **View > Wireframe > Option Box** from the menubar to display the **Wireframe Options** window. Using the options in this window, you can change the color of the wireframe.

Shaded
| **UV Editor menubar:** | View > Shaded |

This tool is used to display the UVs in semi-transparent shading. You can change the color of the shading by choosing **View > Shaded > Option Box** from the **UV Editor** menubar.

UV Distortion
| **UV Editor menubar:** | View > UV Distortion |

This tool is used to display the stretched and compressed UVs. The red faces indicate stretching and the blue faces indicate compression, and white faces indicate optimal UVs. You can also enable this option by choosing **View > UV Distortion** from the **UV Editor**

UV Toolkit
The **UV Toolkit** is located at the right side of **UV Editor** and contains all tools to manipulate the UVs. In the **UV Toolkit**, the tools are arranged in various areas.

Cylindrical
| **Menubar:** | UV > Create > Cylindrical |

The cylindrical mapping technique is used for cylindrical projection of UVs on a polygonal object. This technique works best for objects that can be completely enclosed in the cylindrical projection area. Before applying cylindrical mapping, you need to assign texture to the object.

To understand the cylindrical mapping, create a polygon cylinder in the viewport and press and hold the right mouse button over it; a marking menu will be displayed. Choose **Assign Favorite Material > Lambert** from the marking menu; the **Attribute Editor** will be displayed on the right side of the viewport. In the **Attribute Editor**, choose the checker button on the right of the **Color** attribute in the **lambert#** tab; the **Create Render Node** window will be displayed. Choose the **Checker** button from this window; the checker texture will be assigned to the object. Press 6 to display the checker texture on the object. You will observe that the checker pattern created on the object is in distorted form.

The checker texture helps you to judge how the texture will appear. If the checkers in the checker map stretch, the texture will also stretch. To avoid the texture from stretching, select the cylinder from the viewport and choose **UV > Create > Cylindrical** from the menubar; the cylindrical mapping projection manipulators will be displayed on the object, as shown in Figure 5-2.

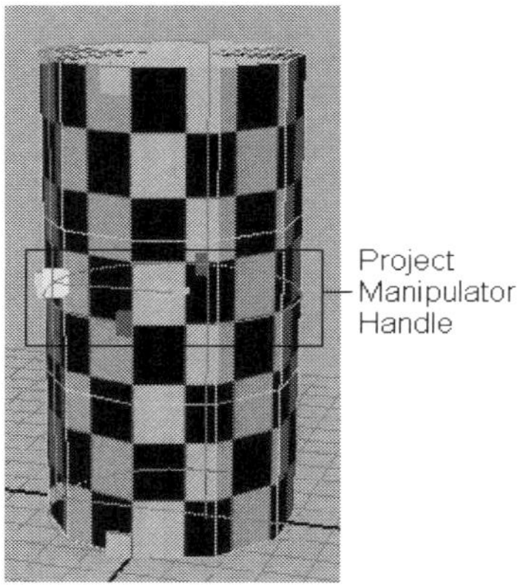

Spherical

| **Menubar:** | UV > Create > Spherical |

The spherical mapping technique creates UVs using a projection based on a spherical shape wrapped around a mesh. This technique works best for spherical objects that can be completely enclosed in a spherical projection area.

Figure 5-2 The cylindrical mapping projection manipulators

Before applying spherical mapping to an object, you need to assign a texture to the object. To do so, press and hold the right mouse button over the object in the viewport; a marking menu will be displayed. Choose **Assign Favorite Material > Lambert** from the marking menu; the **Attribute Editor** will be displayed on the right side of the viewport.

In the **Attribute Editor**, choose the checker button on the right of the **Color** attribute; the **Create Render Node** window will be displayed. Choose the **Checker** button from this window; the checker texture will be assigned to the object. Press 6 to display the checker texture on the object. You will observe that the created checker pattern is in a distorted form.

Next, select the object in the viewport and then choose **UV > Create > Spherical** from the menubar; the spherical mapping projection manipulators will be displayed on the object, as shown in Figure 5-3.

You can adjust these mapping manipulators to set the mapping coordinates. You can also change the default settings of the spherical mapping. To do so, choose **Create > Spherical > Option Box** from the menubar; the **Spherical Mapping Options** window will be displayed, as shown in Figure 5-4. Set the required parameters in the window and then choose the **Project** button.

UV Mapping

Figure 5-3 *The spherical mapping projection manipulators*

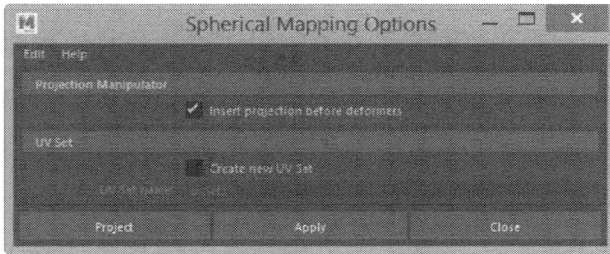

Figure 5-4 *The **Spherical Mapping Options** window*

Camera-Based

Menubar: UV > Create > Camera-Based

The camera-based mapping technique is used to create UV texture for the coordinates on a polygonal object, based on the current camera view. In this type of projection, UVs are created on the object based on faces visible in the view plane. To create UVs using this tool, select a polygonal object from the viewport, press and hold the right-mouse button over it. Next, choose **Face** from the marking menu displayed; the face mode will be activated. Now, select the faces for which you want to create the UVs. After selecting the faces, choose **UV > Create > Camera-Based** from the menubar; the projection will be applied to the selected faces.

Normal-Based

Menubar: UV > Create > Normal-Based

The normal-based mapping technique creates UVs based on the normals of associated vertices. It creates a planar projection based on the average vector of the face normals and the active selection.

Planar

Menubar: UV > Create > Planar

The planar mapping technique is used to map UV texture coordinates on the mesh through an imaginary plane. This is the best suited technique for objects with a flat surface. On applying this projection to an object, the projection manipulator handles will be displayed on that object, as shown in Figure 5-5.

Contour Stretch

Menubar: UV > Create > Contour Stretch

The contour stretch technique is used to project a texture image onto the selected polygons of an object. Contour

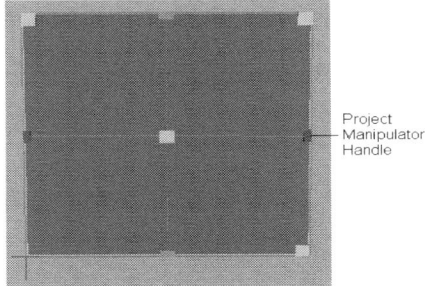

Figure 5-5 *The planar mapping projection manipulator handle*

stretch mapping analyzes a four-cornered selection to determine how to stretch the UV coordinates of the polygons over the image. It does not have the same alignment and positioning options as are available in other mapping methods.

TUTORIAL

All the files used in the tutorial can be downloaded from the CADSoft website (*www.cadsofttech.com*). These files are compressed in zip file format and are required to be extracted before using them in the tutorials. The path of the files is as follows: *Textbooks > Animation and Visual Effects > Maya > Autodesk Maya 2019 for Novices*

Tutorial 1

In this tutorial, you will model a wooden box and then apply texture to it. The final rendered output of the model is displayed in Figure 5-6.   **(Expected time: 15 min)**

Figure 5-6 The final rendered output of the model

The following steps are required to complete this tutorial:

a. Create a project folder.
b. Download the texture file.
c. Create a polygon cube.
d. Fit the texture using the 2D UV coordinates.
e. Change the background color of the scene.
f. Save and render the scene.

Creating a Project Folder

Create a new project folder with the name *c05_tut1* at *\Documents\maya2019* and then save the file with the name *c05tut1* folder, as discussed in Tutorial 1 of Chapter 2.

Downloading the Texture File

In this section, you need to download the texture file.

1. Download the *c05_maya_2019_tut.zip* file from *www.cadsofttech.com*. The path of the file is as follows: *Textbooks > Animation and Visual Effects > Maya > Autodesk Maya 2019 for Novices*.

UV Mapping

2. Extract the contents of the zip file to the *Documents* folder. Next, copy the *woodbox-texture.jpg* file from *\Documents\maya2019\c05_maya_2019_tut* to *\Documents\maya2019\c05_tut1\sourceimages*.

Creating a Polygon Cube

In this section, you need to create a polygon cube.

1. Choose **Create > Objects > Polygon Primitives > Cube** from the menubar and click in the viewport; a cube is created in the persp viewport.

2. In the **Channel Box / Layer Editor**, expand the **polyCube1** node in the **INPUTS** area and then set **8** as the value for the **Width**, **Height**, and **Depth** attributes.

Fitting Texture Using the 2D UV Coordinates

In this section, you need to apply the texture to the polygon cube using the 2D UV coordinates.

1. Choose **Windows > Editors > Rendering Editors > Hypershade** from the menubar; the **Hypershade** window is displayed.

2. Choose the **Lambert** shader from the **Create** panel; a lambert shader node is created in the **Browser** panel with the name **lambert#**. Press and hold the CTRL key and double-click on the **lambert#** shader in the **Browser** panel; the **Rename node** window is displayed, as shown in Figure 5-7. Enter **Wood box** in the **Enter new name** text box and then choose the **OK** button; the **Lambert** shader is renamed to *Wood box*. Also, the **Wood box** tab is displayed in **Property Editor**.

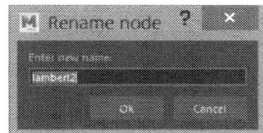

Figure 5-7 The Rename node window

3. In the **Wood box** tab of **Property Editor**, click on the checker button next to the **Color** attribute, as shown in Figure 5-8; the **Create Render Node** window is displayed. In this window, choose the **File** button; the **File Attributes** area is displayed in the **file1** tab of the **Property Editor**. Click on the folder icon located on the right of the **Image Name** attribute; the **Open** dialog box is displayed. In this dialog box, select the **woodbox-texture.jpg** file and then choose the **Open** button.

4. Select the polygon cube in the viewport. In the **Hypershade** window, press and hold the right mouse button over the **Wood box** shader; a marking menu is displayed. Choose the **Assign Material To Selection** option from this marking menu; the texture is applied to the cube. Now, click anywhere in the viewport and press 6 to view the texture in the viewport. Figure 5-9 shows the polygon cube with the texture applied.

5. Make sure the cube is selected and then choose **Windows > Editor > Modeling Editors > UV Editor** from the menubar; the **UV Editor** is displayed. Next, choose **View > Grid** from the **UV Editor** menubar; the grid becomes invisible and the UV shell for the cube is displayed, as shown in Figure 5-10.

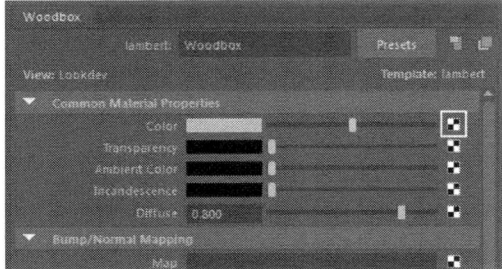

Figure 5-8 Clicking on the checker button next to the **Color** attribute in the **Common Material Properties** panel

Figure 5-9 The wood texture applied to the cube

6. Press and hold the right mouse button in the empty space of the **UV Editor**; a marking menu is displayed. Choose **UV** from the marking menu and select all the UVs. Invoke **Scale Tool** from the Tool Box; various handles are displayed. Scale the selected UVs using the marque selection along the X axis by dragging the red handle, align the edges with the vertical lines of the *woodbox-texture.jpg*. The entire texture is mapped on to the cube, except on to the two areas that are not covered in the V area. Figure 5-11 shows the selected 2D UV texture coordinates after scaling them.

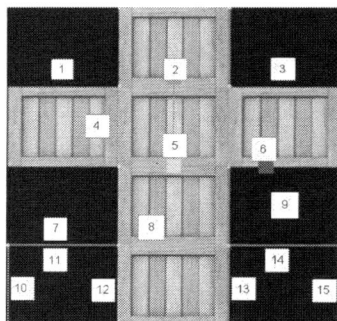

Figure 5-10 The UV shell for cube

Figure 5-11 The 2D UV texture coordinates after scaling

Note
*In the **UV Editor**, the area of the texture within the UV coordinates will only be visible on the object in the viewport.*

7. Press and hold the right mouse button in the empty space of **UV Editor**; a marking menu is displayed. Next, choose **Edge** from the marking menu. Select edge 12 from **UV Editor**, refer to Figure 5-12. Now, choose **Cut/Sew > Cut** from the **UV Editor** menubar; the UVs of selected edges are separated from the edge. Next, select edge 4, refer to Figure 5-11 and choose **Cut/Sew > Move and Sew** from the **UV Editor** menubar; the edge corresponding to the selected edge of the 2D texture coordinate is moved and sewed. Figure 5-13 displays the 2D UV coordinate partially mapped over the texture.

UV Mapping

Figure 5-12 Edge 12 to be selected from the **UV Editor**

Figure 5-13 The 2D UV coordinate partially mapped over the texture

8. In **UV Editor**, select edge 13, refer to Figure 5-18. Choose **Cut/Sew > Cut** from the **UV Editor** menubar; the UVs of selected edges are separated from the edge. Now, select edge 6, refer to Figure 5-18. Choose **Cut/Sew > Move and Sew** from the menubar; the edge corresponding to the selected edge of the 2D texture coordinate is moved and sewed to match the 2D UV coordinate completely with the texture. Figure 5-14 displays the UV coordinate completely mapped over the texture.

9. Close **UV Editor** and the **Hypershade** window. Now, you can rotate the view in the persp viewport to check that the texture is properly applied on the polygon cube, or not. You can also scale the UVs, if the texture is stretched.

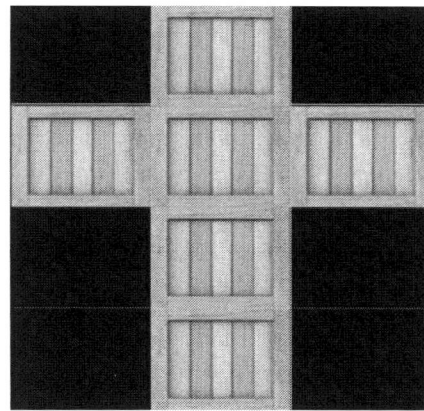

Figure 5-14 The 2D UV coordinate completely mapped over the texture

Changing the Background Color of the Scene

In this section, you need to change the background color of the scene.

1. Choose **Windows > Editors > Outliner** from the menubar; the **Outliner** window is displayed. Click on the **persp** camera in the **Outliner** window; the **perspShape** tab is displayed in the **Attribute Editor**.

2. In the **perspShape** tab, expand the **Environment** area and drag the **Background Color** slider bar toward right to change the background color to white. Close the **Outliner** window.

Saving and Rendering the Scene

In this section, you will save the scene that you have created and then render it. You can view the final rendered image of the model by downloading the *c05_maya_2019_rndr.zip* file from *www.cadsofttech.com*. The path of the file is as follows: *Textbooks > Animation and Visual Effects > Maya > Autodesk Maya 2019 for Novices..*

1. Choose **File > Save Scene** from the menubar.

2. Maximize the persp viewport, if it is not already maximized. Choose the **Render the current frame** button from the Status Line to view the final output.

EXERCISES

The rendered output of the models used in the following exercises can be accessed by downloading the *c05_maya_2019_exr.zip* file from *www.cadsofttech.com*. The path of the file is as follows: *Textbooks > Animation and Visual Effects > Maya > Autodesk Maya 2019 for Novices.*

Exercise 1

Create a model of the interior of a house, as shown in Figure 5-15, and unwrap it.

(Expected time: 20 min)

Figure 5-15 The unwrapped model of the interior of a house

Exercise 2

Create a model of the exterior of a house, as shown in Figure 5-16, and unwrap it.

(Expected time: 20 min)

Figure 5-16 The unwrapped model of the exterior of a house

Chapter 6

Shading and Texturing

Learning Objectives

After completing this chapter, you will be able to:
- *Navigate in the Hypershade window*
- *Use shaders*

INTRODUCTION

In this chapter, you will learn to apply shading and textures. Textures are applied to objects to provide them a realistic appearance.

WORKING IN THE Hypershade WINDOW

menubar:	Windows > Editors > Rendering Editors > Hypershade

The options in the **Hypershade** window can be used to create, edit, and connect the rendering nodes such as textures, materials, and lights. To open this window, choose **Windows > Editors > Rendering Editors > Hypershade** from the menubar; the **Hypershade** window will be displayed, as shown in Figure 6-1.

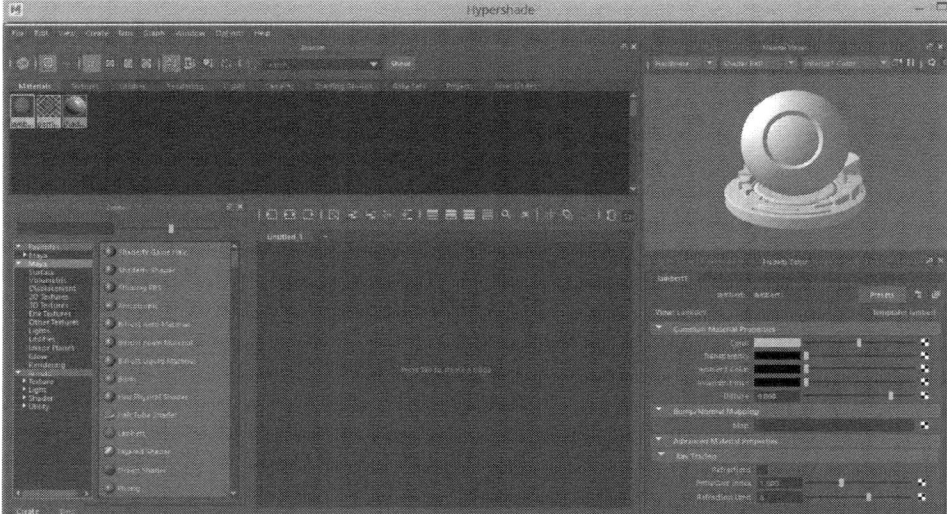

Figure 6-1 The **Hypershade** window

Browser Panel

The top panel of the **Hypershade** window contains ten tabs that are used to access rendering components, refer to Figure 6-2. These tabs also correspond to various objects present in the viewport. For example, the **Materials** tab contains all materials that have been used in the scene and the **Lights** tab contains lights that are added to the scene, and so on.

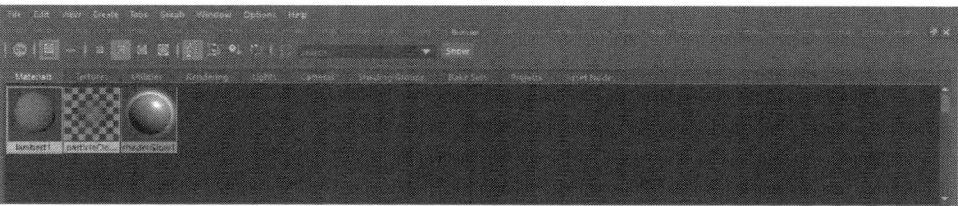

Figure 6-2 The **Browser** tab in the **Hypershade** window

Shading and Texturing

Browser Panel Toolbar

The Browser panel toolbar is located on the top of the **Hypershade** window, refer to Figure 6-3. The buttons in this toolbar are used to control viewing, listing, and ordering of the options in the different tabs of the **Browser** panel.

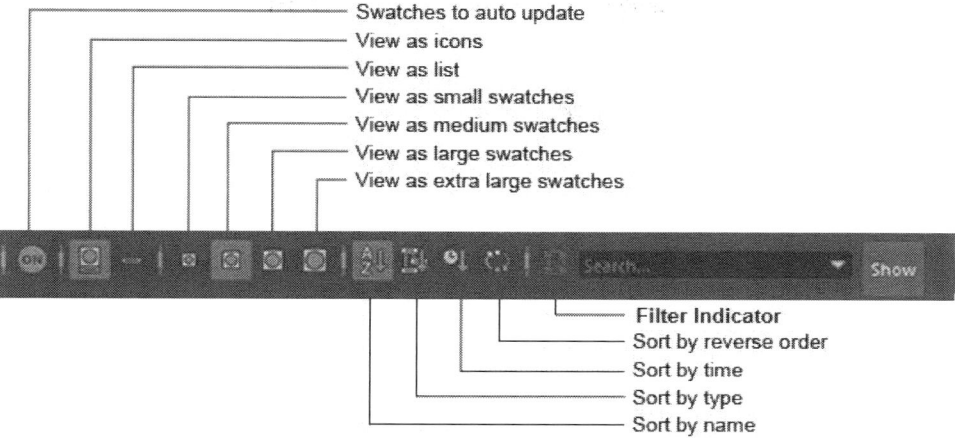

Figure 6-3 The Browser panel toolbar in the **Hypershade** window

Work Area

This area is located on the right of the **Create** panel. By default, the **Untitled_1** tab is displayed in this panel, refer to Figure 6-1. The Work Area displays the shading network for the selected node. A shading network is an arrangement of nodes that affect the final look of the surface on which the material is applied.

A toolbar, referred to as the Work Area toolbar, is located on the top of the Work Area. It consists of various buttons that are used to control shading and texturing, refer to Figure 6-4.

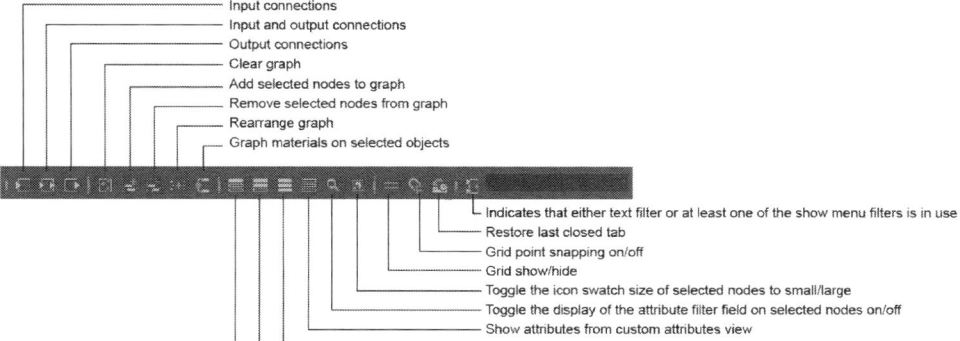

Figure 6-4 The main toolbar of the **Hypershade** window

PROPERTY EDITOR

In Maya, the shaders are controlled by attributes. To view these attributes, click on a shader in the **Browser** panel of the **Hypershade** window; all attributes of the corresponding shader will be displayed in the **Property Editor**, refer to Figure 6-5. You can tear off the **Property Editor** panel from the **Hypershade** window. By choosing top right button you can toggle between **Lookdev view** and **Attribute Editor view**.

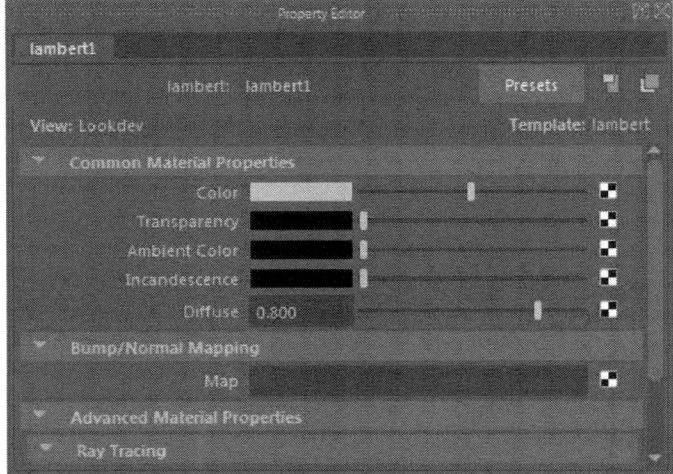

Figure 6-5 The **Property Editor** *displaying the Lambert shader attributes*

The **File** button allows you to add images as maps and textures, whereas the **PSD File** button allows you to add the Photoshop file as maps and textures. If you choose the **File** button from the **Create Render Node** window, the **File Attributes** area will be displayed in the **Property Editor**. Choose the folder button on the right of the **Image Name** attribute; the **Open** window will be displayed. Choose the image file from the location on the disk and then choose the **Open** button. Similarly, add the PSD texture by choosing the **PSD File** button from the **Create Render Node** window.

Bump/Normal Mapping

The **Bump Mapping/Normal Mapping** area is used to add bump effect to an object on rendering. To make this attribute visible, choose the **Toggle between Lookdev view and Attribute Editor view** button in the Property Editor. This attribute does not modify the surface of the object, but it shows roughness on the surface on rendering. To apply bump map to an object, choose the checker button on the right of the **Map** attribute; the **Create Render Node** window will be displayed. Select the map or texture to which you want to apply the bump and then choose the **Close** button. Render the object to see the bump effect. Figure 6-6 shows the object after applying different textures to the **Map** attribute.

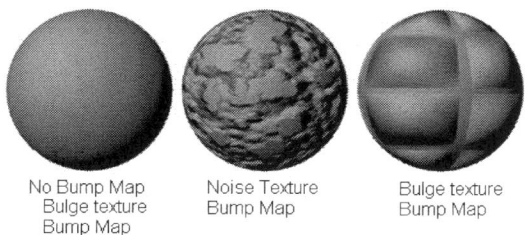

Figure 6-6 *Object after applying different textures to the* ***Bump Mapping*** *attribute*

Special Effects

The options in this area are used to set the parameters of special effects applied to an object. These special effects are visible only when the object is rendered. This area consists of only the **Glow Intensity** attribute to add glow effect on the edges of objects. To display the **Glow Intensity** attribute, choose the **Toggle between Lookdev view and Attribute Editor view** button available on the top-right corner of the **Property Editor**. The glow effect is discussed next.

Glow Intensity

The **Glow Intensity** attribute is used to add glow to the edges of an object, as shown in Figure 6-7. To add glow intensity to an object, enter the required value in the **Glow Intensity** edit box located in the **Special Effects** area, or drag the slider on the right of the **Glow Intensity** attribute. Next, set the renderer to **Maya Software** and then choose the **Render the current frame** button from the Status Line to render and adjust the glow as required. You can also hide the source of the glow object. To do so, select the **Hide Source** check box from the **Special Effects** area. You can also add a light glow source to an object. To do so, choose the checker button on the right of the **Glow Intensity** attribute; the **Create Render Node** window will be displayed. Select the **Glow** option on the left pane of the window and then select the **Optical FX** option from the right pane of the window, as shown in Figure 6-8. The **Optical FX** option will be added to the object. Now, render the scene to see the final effect.

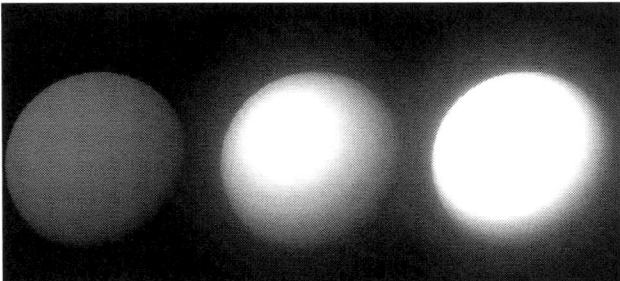

Figure 6-7 *Spheres with different glow intensities*

EXPLORING THE SHADERS

As you know, that the **Create** panel is located on the bottom-left in the **Hypershade** window. This panel has different types of nodes which are used to create different shading networks. These nodes are divided into three categories: **Favorites**, **Maya**, and **Arnold**. These categories are further divided into sections. Among these sections, the **Surface** section which comes under the **Maya** category consists of all shaders/nodes that are required to apply texture to an object. The **Surface** section is discussed next.

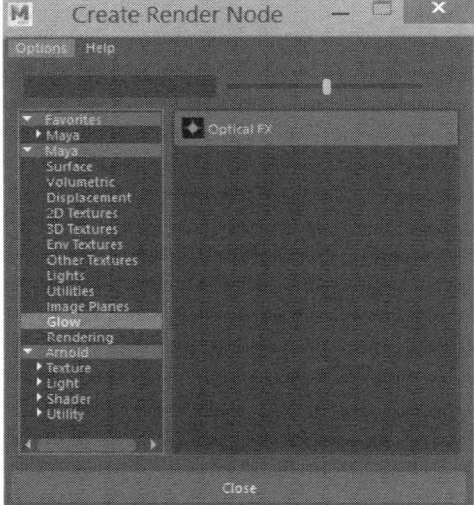

*Figure 6-8 The **Optical FX** option displayed*

Surface

By default, all shaders/nodes of this section are displayed in this section. The **Surface** section is mainly used to define the physical appearances of objects. The most commonly used shaders in the **Surface** section are discussed next.

Anisotropic

The **Anisotropic** shader is used to create deformed surfaces such as foil wrapper, wrapped plastic, hair, or brushed metal. The directions of the highlights change according to direction of the object in the viewport. Due to this property, the elliptical or anisotropic highlights are created, as shown in Figure 6-9. Some of the examples of the objects created by applying the **Anisotropic** shader are CDs, feather, and utensils.

Bifrost Aero Material

The **Bifrost Aero Material** shader is a mental ray material that creates atmospheric effects such as smoke and mist. This shader gets automatically applied to the aero and bifrost Aero Mesh objects while creating a Bifrost simulation. It is a volume ray

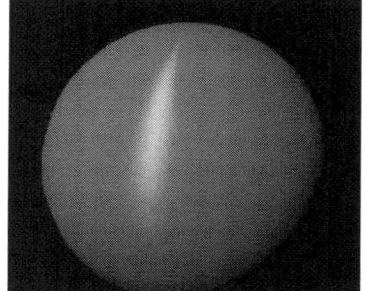

*Figure 6-9 The **Anisotropic** shader applied to an object*

marcher that accumulates the shading contributions from emission, absorption, and scattering at each step based on the density. You need to install **mental ray** renderer if you want to use it. **Maya Software**, **Maya Hardware**, and **Maya Vector** renderers do not support this shader. Some of its attributes are supported by the **Maya Hardware 2.0** renderer.

Shading and Texturing

Bifrost Foam Material

The **Bifrost Foam Material** shader is used to create bubbles, foam, and spray effects. By using this shader, you can also generate foam if you emit liquid into an existing liquid with a different density, such as in the case of a hot-tub liquid effect.

Blinn

The **Blinn** shader is mainly used to create shiny metallic surfaces such as brass and aluminium. Figure 6-10 shows the **Blinn** shader applied to a sphere.

Lambert

The **Lambert** shader is mainly used to create unpolished surfaces. This shader diffuses and scatters light evenly on the object created in the viewport, thus giving it an unpolished appearance. It has no specular highlighting properties. Figure 6-11 shows a sphere with the **Lambert** shader applied to it.

*Figure 6-10 The **Blinn** shader applied to a sphere*

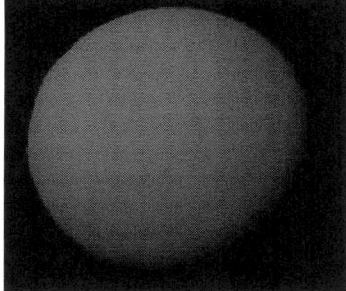

*Figure 6-11 The **Lambert** shader applied to a sphere*

Layered Shader

The **Layered Shader** is used when multiple materials are needed to be applied to the surface of an object. Figure 6-12 shows an object with the **Layered Shader** applied to it. It helps in creating a surface with distinct look and style. In this shader, different textures and shades are blended together to give a realistic look to the surface of an object. The **Layered Shader** takes more time in rendering.

To apply the **Layered Shader**, choose **Windows > Editors > Rendering Editors > Hypershade** from the menubar; the **Hypershade** window will be displayed. In the **Hypershade** window, choose the **Layered Shader** from the left of the **Create** panel; **layeredShader1** will be created in the **Untitled_1** tab. Next, choose the **Lambert** and **Anisotropic** shaders from the **Create** area; the **lambert2** and **anisotropic1** shaders will be created in the **Untitled_1** tab. Click on the **layeredShader1** shader in the **Untitled_1** tab; the **layeredShader1** tab will be displayed in the **Property Editor**, as shown in Figure 6-13. Next, press and hold

*Figure 6-12 The **Layered Shader** applied to an object*

the middle mouse button over the **lambert2** shader in the **Hypershade** window and drag it to the green swatch in the **Layered Shader Attributes** area of the **Attribute Editor**; the **lambert2** swatch is created in the **Layered Shader Attributes** area.

Figure 6-13 The layeredShader1 tab in the Property Editor

Ocean Shader

The **Ocean Shader** is used to create realistic ocean. It can also be used to stimulate waves in the viewport. To use this shader, create a plane in the viewport with the **Width Subdivisions** and **Height Subdivisions** set to **20** each. Next, choose **Windows > Editors > Rendering Editors > Hypershade** from the menubar; the **Hypershade** window will be displayed. Choose **Ocean Shader** from the **Hypershade** window; the **oceanShader1** will be created in the **Browser** area. Select the plane in the viewport, and press and hold the right mouse button over the **OceanShader1** in the **Hypershade** window; a marking menu will be displayed. Choose **Assign Material To Selection** from the marking menu; the material will be applied to the plane in the viewport. Set the renderer to **Maya Software** and then Choose the **Render the current frame** button from the Status Line to render the scene; the plane rendered using **Ocean Shader** is shown in Figure 6-14.

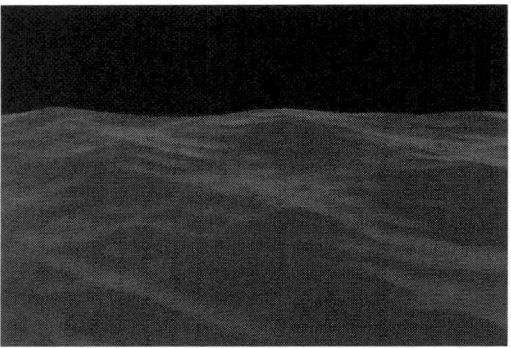

*Figure 6-14 The plane rendered using **Ocean Shader***

Note

*If you apply the **Ocean Shader** to an object and then choose the **Play forwards** button, you will notice that the in-built animation is being played in the viewport. Also, while using the **Ocean Shader**, you always need to apply general lighting to brighten the scene.*

Shading and Texturing

Phong

The **Phong** shader is used to add shine to an object, as shown in Figure 6-15. A phong surface reflects light, thus creating a specular highlight on the object. The **Phong** shader has certain characteristics such as diffusion and specularity that can be used to create smooth light reflecting surfaces. For example, you can create plastics, glass, ceramics, and most of the metals by using the **Phong** shader.

Phong E

The **Phong E** shader is used to produce glossy surfaces. This shader is perfect for creating plastics, bathroom accessories, and car modeling. Figure 6-16 shows the **Phong E** shader applied to a sphere.

Figure 6-15 The **Phong** shader applied to a sphere

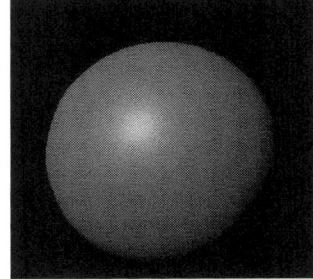

Figure 6-16 The **Phong E** shader applied to a sphere

Ramp Shader

The **Ramp Shader** is used to apply additional control over the colors of shader with respect to change in light and direction of the object in the viewport. All attributes related to colors in this shader are controlled by ramps. Ramps are known as gradients and are used to create smooth transitions among different colors. You can apply the **Ramp Shader** to an object in the viewport. To do so, invoke the **Hypershade** window and choose the **Ramp Shader** from the **Create** panel. Next, click on the **rampShader1** shader in the **Untitled_1** tab; the attributes of the **rampShader1** will be displayed in the **rampShader1** tab in the **Property Editor**, as shown in Figure 6-17 and then click on the color ramp on the right of the **Selected Color** attribute; a new color entry will be created. Drag the circular handle on top of the new color node to adjust it, as shown in Figure 6-18.

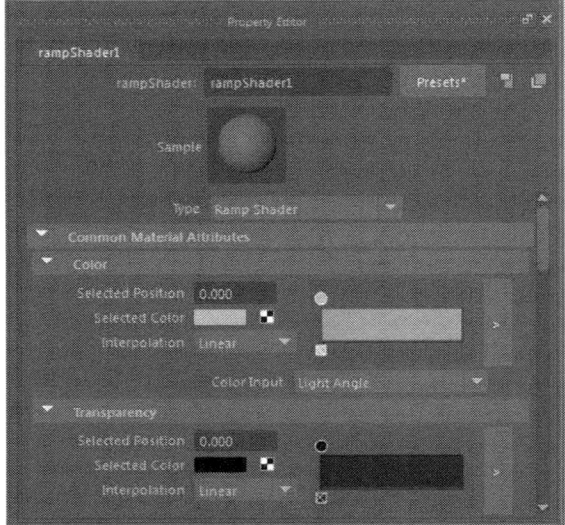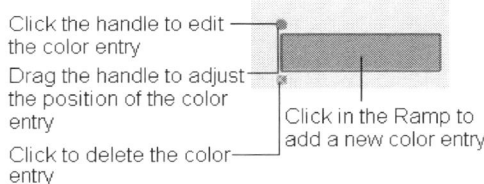

*Figure 6-17 The **rampShader1** tab in the **Property Editor***

*Figure 6-18 The color ramp in the **Color** area of the **rampShader1***

Select the circular handle and choose the color swatch on the right of the **Selected Color** attribute. To add a map to a particular color entry, select the handle and choose the checker box on the right of the **Selected Color** attribute; the **Create Render Node** window will be displayed. In the **Create Render Node** window, choose the **Mountain** texture and then select the object in the viewport. Next, press and hold the right mouse button on the **rampShader1** in the **Hypershade** window and choose the **Assign Material To Viewport Selection** option from the marking menu; the object after applying the **rampShader1** will appear.

 Note
*You can also assign different color effects to an object by changing the values of the **Interpolation** and **Color Input** attributes in the **rampShader1** tab of the **Property Editor**.*

Shading and Texturing

TUTORIALS

All the files used in the tutorials can be downloaded from the CADSoft website (*www.cadsofttech.com*). These files are compressed in zip file format and are required to be extracted before using them in the tutorials. The path of the files is as follows: *Textbooks > Animation and Visual Effects > Maya > Autodesk Maya 2019 for Novices*

Tutorial 1

In this tutorial, you will create a polygon cube and apply texture of an old house to it, refer to Figure 6-19. **(Expected time: 30 min)**

Figure 6-19 *The textured model of the cube*

The following steps are required to complete this tutorial:

a. Create a project folder.
b. Download texture files.
c. Create a polygon cube.
d. Apply the checker pattern to the cube.
e. Create a texture in Adobe Photoshop.
f. Apply the texture to the cube.
g. Change the background color of the scene.
h. Save and render the scene.

Creating a Project Folder

Create a new project folder with the name *c06_tut1* at *\Documents\maya2019* and then save the file with the name *c06tut1* as discussed in Tutorial 1 of Chapter 2.

Downloading Texture Files

In this section, you will download the texture files.

1. Download the *c06_maya_2019_tut.zip* file from *www.cadsofttech.com*. The path of the file is as follows: *Textbooks > Animation and Visual Effects > Maya > Autodesk Maya 2019 for Novices*.

2. Extract the contents of the zip file to the *Documents* folder. Navigate to *\Documents\ c06_maya_2019\tut* and then copy the entire texture files to the *\Documents\maya2019\c06_tut1\ sourceimages*.

Creating a Polygon Cube

In this section, you will create a cube using cube polygon primitive.

1. Choose **Create > Objects > Polygon Primitives** from the menubar; a flyout is displayed. Next, choose the **Interactive Creation** option from the flyout.

2. Choose **Create > Objects > Polygon Primitives > Cube > Option Box** from the menubar; the **Tool Settings (Polygon Cube Tool)** panel is displayed. Alternatively, double-click on the **Polygon Cube** icon in the **Polygons** tab of the Shelf to display the **Tool Settings (Polygon Cube Tool)** panel.

3. Enter values in this window, as shown in Figure 6-20. Next, click in the persp viewport; a cube is displayed in the persp viewport.

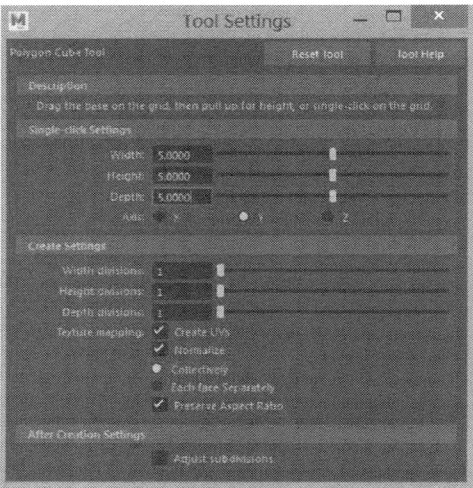

Figure 6-20 The **Tool Settings (Polygon Cube Tool)** *panel*

Applying the Checker Pattern to the Cube

In this section, you will apply the checker pattern to the cube.

1. Choose **Windows > Editors > Rendering Editors > Hypershade** from the menubar; the **Hypershade** window is displayed.

2. Choose the **Lambert** shader from the **Create** panel in the **Hypershade** window; the **Lambert** shader with the name **lambert2** is created in the **Browser** panel. Next, press the CTRL key

Shading and Texturing 6-13

and double-click on the **lambert2** shader in the **Browser** panel; the **Rename node** dialog box is displayed. Enter **initial_texture** in the **Enter new name** edit box, and then choose the **OK** button; the **lambert2** shader is renamed as *initial_texture*.

3. Click on the **initial_texture** shader in the **Browser** panel; the **initial_texture** tab is displayed in the **Property Editor**.

4. In the **Common Material Properties** area of the **initial_texture** tab, click on the checker button corresponding to the **Color** attribute, refer to Figure 6-21; the **Create Render Node** window is displayed.

5. Choose the **Checker** button from the **Create Render Node** window, refer to Figure 6-22.

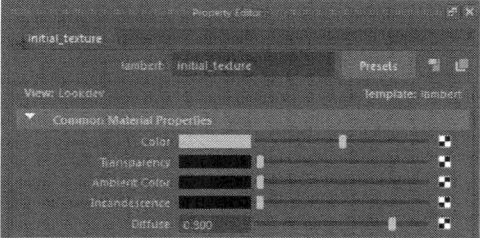

Figure 6-21 Choosing the checker button corresponding to the **Color** attribute

6. Select the polygon cube in the viewport. Now, in the **Untitled_1** tab of the **Hypershade** window, press and hold the right mouse button on the **initial_texture** shader; a marking menu is displayed. Choose the **Assign Material To Selection** option from the marking menu, as shown in Figure 6-23; the *initial_texture* shader is applied to the polygon cube. Press 6; the texture is displayed on the cube in the viewport.

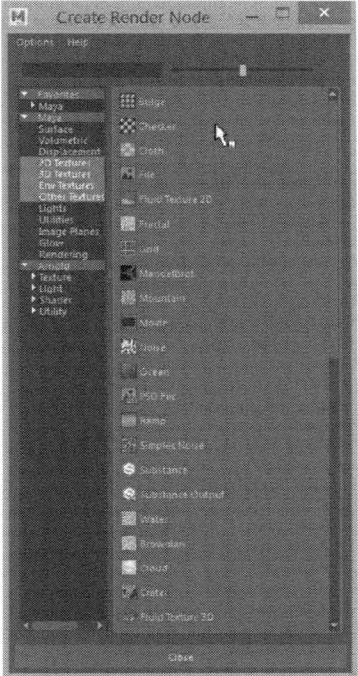

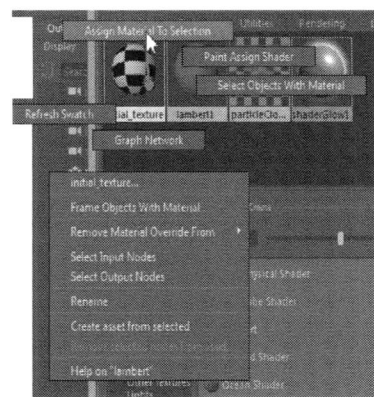

Figure 6-22 Choosing the **Checker** button from the **Create Render Node** window

Figure 6-23 Choosing **Assign Material To Selection** from the marking menu

7. Make sure the cube is selected in the viewport. Select the **Modeling** menuset from the **Menuset** drop-down list if not already selected. Choose **UV > UV Editor** from the menubar; the **UV Editor** is displayed, as shown in Figure 6-24.

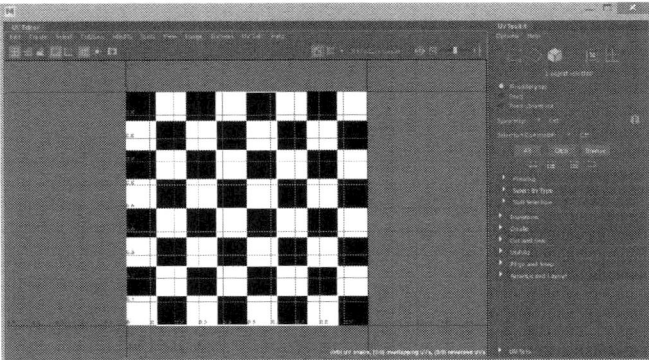

Figure 6-24 The UV Editor

8. In the **UV Editor**, choose the **Texture borders** button if not already chosen; the uvs of the cube are highlighted. Now, choose **Image > UV Snapshot** from the **UV Editor** menubar; the **UV Snapshot Options** window is displayed, as shown in Figure 6-25. Choose the **Browse** button; the **Save Snapshot** dialog box is displayed. In this dialog box, browse to the location *\Documents\maya2019\c06_tut1\images*. Next, save the UV snapshot with the name **UV snapshot** and choose the **Save** button; the **Save Snapshot** dialog box closes. Next, enter **1024** in the **Size X (px)** edit box in the **UV snapshot** area; you will notice that **1024** gets automatically entered in the **Size Y (px)** edit box. Choose the **Apply and Close** button from the **UV Snapshot Options** window. Close the **Hypershade** and the **UV Editor** windows.

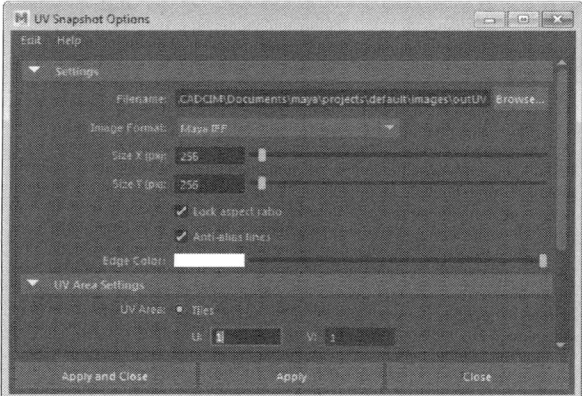

Figure 6-25 The UV Snapshot Options window

Now, you will open the *UV snapshot.iff* file in Adobe Photoshop.

Creating a Texture in Adobe Photoshop

In this section, you will create a texture for the cube using Adobe Photoshop.

Shading and Texturing

1. Open the *UV snapshot.iff* file in Adobe Photoshop. The file opens in the canvas area and a layer with the name **Layer 0** is created in the **Layers** panel.

2. Choose the **Create a new layer** button in the **Layers** panel; a new layer with the name **Layer 1** is created.

3. Make sure the newly created layer is selected and set **Set foreground color** to black color in the Tool Box. Next, press ALT+BACKSPACE; the **Layer 1** is filled with black color.

4. Move this layer below the **Layer 0**; the faces are now visible in the canvas area, as shown in Figure 6-26.

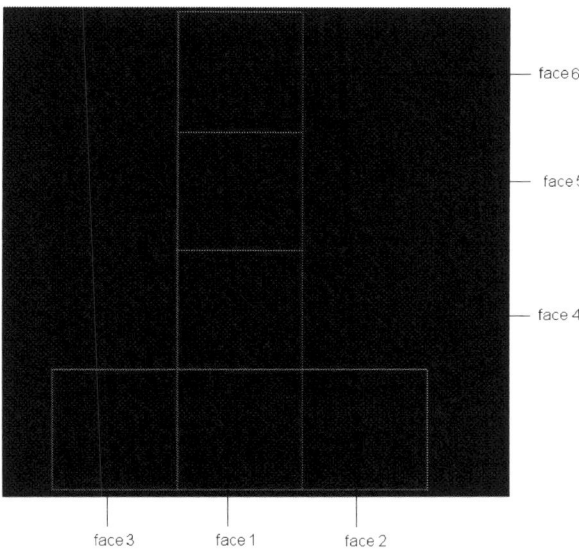

Figure 6-26 UVs in the canvas area of Photoshop

5. Choose **File > Open** from the menubar; the **Open** dialog box is displayed. In this dialog box, browse to *\Documents\maya2019\c06_tut1\sourceimages\frontwalltexture.jpg* and choose the **Open** button; the *frontwalltexture.jpg* is loaded. Choose **Move Tool**, and drag the image and place it on face 1, refer to Figure 6-27. Press CTRL+T; **Transform Tool** is activated. Next, scale the image such that it fits into face 1, as shown in Figure 6-27. Next, press ENTER; the transformation is applied.

6. Choose **Burn Tool** and darken **Layer 2**.

7. Create a new layer, and using **Brush Tool**, create different patterns to make the image dirty with opacity equal to 15 and brush size equal to 5, as shown in Figure 6-28.

8. Open the files *doortexture.jpg* and *windowtexture.jpg* from the *sourceimages* folder, as discussed earlier. Next, choose **Move Tool** and place the images on face 1. Invoke **Transform Tool** by pressing CTRL+T, and scale the textures to fit them on face 1, as shown in Figure 6-29.

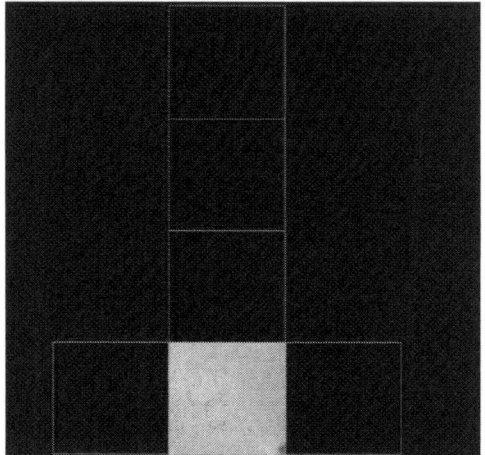

Figure 6-27 Fitting the image into face 1

Figure 6-28 Different patterns created on the image

Figure 6-29 The textures placed on face 1

9. Select the layer having door, and then choose the **Add a layer style** button from the **Layers** panel; a flyout is displayed. Choose the **Bevel Emboss** option from the flyout; the **Layer Style** dialog box is displayed. In this dialog box, enter the values, as shown in Figure 6-30. Now, choose the **OK** button, a depth is created in the door. Repeat the same procedure to create depth in the window.

Shading and Texturing

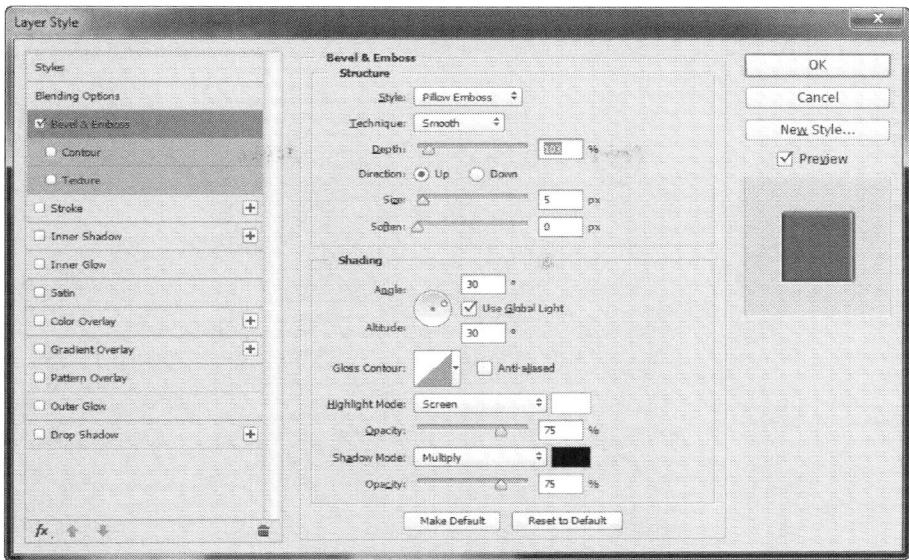

*Figure 6-30 The **Layer Style** dialog box*

10. Select the layer having window in the canvas and press and hold ALT, and then drag the layer; a duplicate copy of the window is created. Next, place the window on face 1, as shown in Figure 6-31.

11. Open the *sidewallstexture.jpeg* file from the *sourceimages* folder, as discussed earlier, and place it on face 2 and face 3. Create different patterns on the faces using **Burn Tool** and **Brush Tool**, as shown in Figure 6-32.

Figure 6-31 A copy of window created on face 1   *Figure 6-32 Patterns created on face 2 and face 3*

12. Similarly, apply the *roof.jpeg, backside.jpeg,* and *ground.jpeg* texture files at the top, back, and base of the cube, respectively, as shown in Figure 6-33. Next, create different patterns on the textures to make the textures worn out. Make the area below the windows darker to show seepage in the walls. Next, turn off **Layer 0** so that the seams are not visible in the texture.

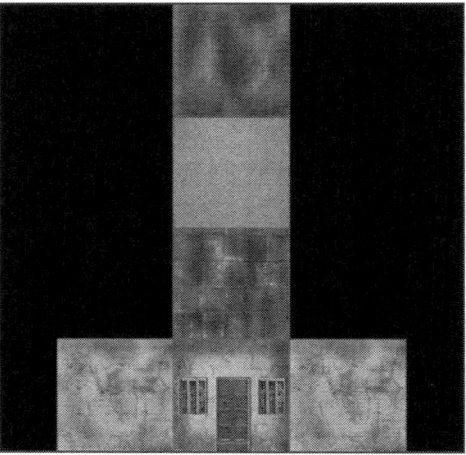

Figure 6-33 Patterns created on all faces

13. Choose **File > Save As** from the menubar; the **Save As** dialog box is displayed. In this dialog box, enter **Cube_UVs** in the **File Name** text box. Next, make sure the **Photoshop (*.PSD;*.PDD)** option is selected in the **Format** drop-down list. Next, browse to *\Documents\ maya2019\c06_tut1\sourceimages* and choose the **Save** button; the file is saved at the specified location.

Next, you will switch back to Autodesk Maya and apply the texture created in Photoshop to the cube.

Applying the Texture to the Cube

In this section, you will apply the texture created in Photoshop to the cube.

1. Make sure the cube is selected in the viewport, choose **Windows > Editors > Rendering Editors > Hypershade** from the menubar; the **Hypershade** window is displayed.

2. Choose the **Lambert** shader from the **Create** panel; the **Lambert** shader with the name **lambert3** is created in the **Untitled_1** tab. Press CTRL and then double-click on the **lambert3** shader in the **Create** panel; the **Rename node** window is displayed. Enter **oldhouse** in the text box and press ENTER; the **lambert3** shader is renamed as *oldhouse*. Click on the **oldhouse** shader; the **oldhouse** tab is displayed in the **Property Editor**.

3. In the **oldhouse** tab, click on the checker button corresponding to the **Color** attribute in the **Common Material Properties** area; the **Create Render Node** window is displayed. Choose the **PSD File** button from the **Create Render Node** window; the **psdFileTex1** tab is displayed in the **Property Editor**, as shown in Figure 6-34.

Shading and Texturing

4. Click on the folder icon on the right of the **Image Name** text box in the **File Attributes** area; the **Open** dialog box is displayed. Next, browse and select the **Cube_UVs.psd** and then choose the **Open** button.

5. Select the cube in the persp viewport. In the **Browser** panel of the **Hypershade** window, press and hold the right mouse button over the **oldhouse** shader; a marking menu is displayed. Choose the **Assign Material To Selection** option from the marking menu; the texture is applied to all sides of the cube, as shown in Figure 6-35. Close the **Hypershade** window.

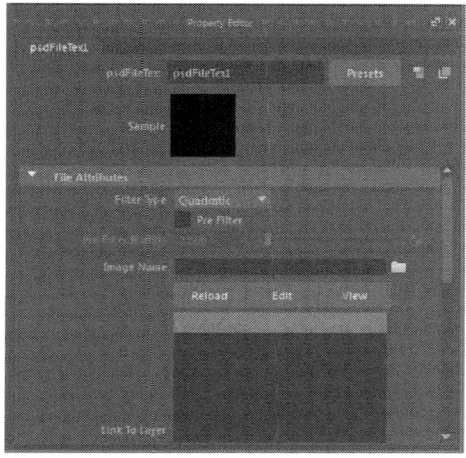

Figure 6-34 The psdFileTex1 tab

Figure 6-35 The texture applied to all sides of the cube

Changing the Background Color of the Scene

In this section, you will change the background color of the scene.

1. Choose **Windows > Editors > Outliner** from the menubar; the **Outliner** window is displayed. Click on the **persp** camera in the **Outliner** window; the **perspShape** tab is displayed in the **Attribute Editor**.

2. Expand the **Environment** area in the **perspShape** tab and drag the **Background Color** slider bar toward right to change the background color to white.

Saving and Rendering the Scene

In this section, you will save the scene that you have created and then render it. You can view the final rendered image of the scene by downloading the *c06_maya_2019_rndr.zip* file from *www.cadsofttech.com*. The path of the file is as follows: *Textbooks > Animation and Visual Effects > Maya > Autodesk Maya 2019 for Novices*.

1. Choose **File > Save Scene** from the menubar to save the scene.

2. For rendering the scene, refer to Tutorial 1 of Chapter 2. The final rendered output is shown in Figure 12-19.

Tutorial 2

In this tutorial, you will create the model of an eyeball and then apply texture to it, as shown in Figure 6-36. **(Expected time: 15 min)**

The following steps are required to complete this tutorial:

a. Create a project folder.
b. Create the NURBS sphere.
c. Assign material to the sphere.
d. Change the background color of the scene.
e. Save and render the scene.

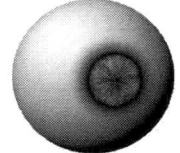

Figure 6-36 Model of an eyeball

Creating a Project Folder

Create a new project folder with the name *c06_tut2* at *\Documents\ maya2019* and then save the file with the name *c06tut2*, as discussed in Tutorial 1 of Chapter 2.

Creating the NURBS Sphere

In this section, you will create the NURBS sphere for the eyeball.

1. Maximize the front-Z viewport and then choose **Create > Objects > NURBS Primitives > Sphere** from the menubar. Next, create a NURBS sphere in the viewport and then set the following parameters in the **Channel Box/Layer Editor**.

   Radius : **2**   Rotate X : **90**   Rotate Z : **-90**

2. Maximize the persp viewport.

Assigning Material to the Sphere

In this section, you will create a material for the eyeball and then assign it to the NURBS sphere.

1. Choose **Windows > Editors > Rendering Editors > Hypershade** from the menubar; the **Hypershade** window is displayed. Select the **Blinn** shader from the **Create** area in this window; the **blinn1** shader is created in the **Browser** panel of the **Hypershade** window.

2. Press and hold the CTRL key and double-click on the **blinn1** shader in the **Browser** panel; the **Rename node** window is displayed. Enter **eye** in the **Enter new name** text box and then choose the **OK** button; the **blinn1** shader is renamed as **eye**.

3. Select the sphere in the viewport. Press and hold the right mouse button on the **eye** shader in the **Browser** panel of the **Hypershade** window and choose **Assign Material to Selection** from the marking menu; the *eye* shader is applied to the sphere.

4. Click on the *eye* shader in the **Hypershade** window; the **eye** tab is displayed in the **Property Editor**.

5. In the **Common Material Properties** area of the **eye** tab, choose the checker button on

Shading and Texturing   6-21

the right of the **Color** attribute; the **Create Render Node** window is displayed. Choose the **Ramp** button from the **Create Render Node** window; the **ramp1** shader tab is created in the **Property Editor**.

6. In the **ramp1** shader tab, select the **U Ramp** option from the **Type** drop-down list and **Bump** from the **Interpolation** drop-down list in the **Ramp Attributes** area. Next, press 6 to view the texture in the viewport.

By default, two color nodes are available in the ramp color area. You will create two more nodes by following the steps given next.

7. Click on the ramp color area twice in the **Ramp Attributes** area; two more nodes are created. Next, arrange the nodes, as shown in Figure 6-37.

8. Select the color node 1, refer to Figure 6-38, from the **Ramp Attributes** area in the **ramp1** tab and then click on the color swatch of the **Selected Color** attribute; the **Color History** palette is displayed. Make sure that the **HSV** option is selected in the drop-down list below the color wheel in the **Ramp Attributes** area. Next, make sure the **HSV** values in the **Color History** palette are as follows:

H: 0   S: 0   V: 0

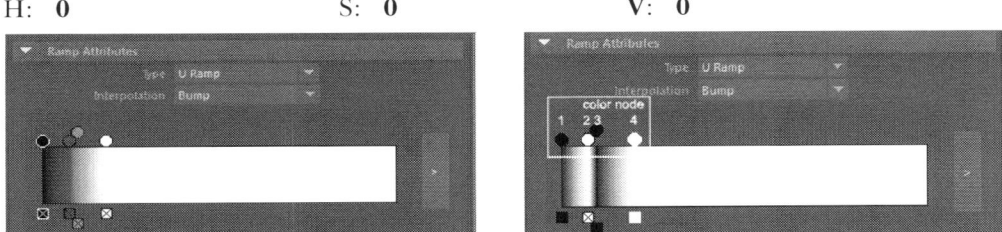

Figure 6-37 The nodes in the color area of the **ramp1** shader

Figure 6-38 The color nodes arranged in the color area

9. Select the color node 2 from the **Ramp Attributes** area in the **ramp1** tab and then click on the color swatch in the **Selected Color** attribute; the **Color History** palette is displayed. Next, enter the following **HSV** values in the **Color History** palette:

H: 0   S: 0   V: 1

10. Select the color node 3 from the **Ramp Attributes** area in the **ramp1** tab and then click on the color swatch in the **Selected Color** attribute; the **Color History** palette is displayed. Next, enter the following **HSV** values in the **Color History** palette:

H: 0   S: 0   V: 0

11. Select the color node 4 from the **Ramp Attributes** area in the **ramp1** tab and then click on the color swatch in the **Selected Color** attribute; the **Color History** palette is displayed. Next, make sure the **HSV** values in the **Color History** palette are as follows:

H: 0   S: 0   V: 1

Figure 6-45 shows the nodes after the colors are assigned and are arranged in the color area. Figure 6-39 displays the eyeball after material is applied to it.

12. Select the color node 2 from the **Ramp Attributes** area in the **Property Editor**, refer to Figure 6-38. Next, click on the checker button on the right of the **Selected Color** attribute; the **Create Render Node** window is displayed. Choose **Fractal** from the **Create Render Node** window, as shown in Figure 6-40; the fractal texture is applied to the eyeball, as shown in Figure 6-41. Also, the **fractal1** tab is displayed in the **Property Editor**.

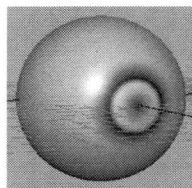

Figure 6-39 Initial eyeball material applied to the NURBS sphere

Figure 6-40 Choosing **Fractal** from the **Create Render Node** window

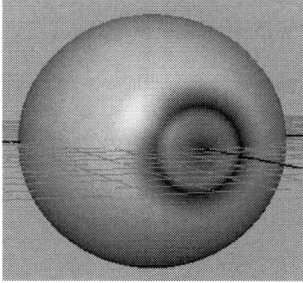

Figure 6-41 Eyeball after applying the fractal texture

13. Make sure the **fractal1** tab is selected in the **Property Editor**. Expand the **Color Balance** area and choose the gray color swatch on the right of the **Default Color** attribute; the **Color History** palette is displayed. In the **Color History** palette, set the following values for **H**, **S**, and **V**:

    H: **199**   S: **0.967**   V: **0.779**

14. In the **Color Balance** area, choose the color swatch corresponding to the **Color Gain** attribute; the **Color History** palette is displayed. In the **Color History** palette, set the following values of **H**, **S**, and **V**:

    H: **199**   S: **0.8**   V: **1**

Shading and Texturing 6-23

15. In the **Color Balance** area, choose the color swatch corresponding to the **Color Offset** attribute; the **Color History** palette is displayed. In the **Color History** palette, set the following values of **H**, **S**, and **V**:
    H: 191   S: 0.967   V: 0.3

16. In the **Common Material Properties** area, set the value of the **Diffuse** and **Eccentricity** attributes as **1** and **0** respectively in the **Specular Shading** area.

Changing the Background Color of the Scene

In this section, you will change the background color of the scene.

1. Choose **Windows > Editors > Outliner** from the menubar; the **Outliner** window is displayed. Select the **persp** camera in the **Outliner** window; the **perspShape** tab is displayed in the **Attribute Editor**.

2. Expand the **Environment** area in the **perspShape** tab and drag the **Background Color** slider bar toward right to change the background color to white.

Saving and Rendering the Scene

In this section, you will save and then render the scene that you have created. You can view the final rendered image of the scene by downloading the *c06_maya_2019_rndr.zip* file from *www.cadsofttech.com*. The path of the file is as follows: *Textbooks > Animation and Visual Effects > Maya > Autodesk Maya 2019 for Novices*.

1. Choose **File > Save Scene** from the menubar to save thse scene.

2. For rendering the scene, refer to Tutorial 1 of Chapter 2. This window shows the final output of the scene, refer to Figure 6-43.

EXERCISES

The rendered output of the models used in the following exercises can be accessed by downloading the *c06_maya_2019_exr.zip* file from *www.cadsofttech.com*. The path of the file is as follows: *Textbooks > Animation and Visual Effects > Maya > Autodesk Maya 2019 for Novices*.

Exercise 1

Create the model of a house, shown in Figure 6-42. Unwrap it and then apply textures to it to get the final output, as shown in Figure 6-43. **(Expected time: 30 min)**

Figure 6-42 The house model before applying the textures

Figure 6-43 The house model after applying the textures

Shading and Texturing

Exercise 2

Create the model of a house, shown in Figure 6-44, and then apply textures to it to get the final output, as shown in Figure 6-45. **(Expected time: 30 min)**

Figure 6-44 *The house model before applying the textures*

Figure 6-45 *The house model after applying the textures*

Exercise 3

Create a scene showing a model of the study table with objects, as shown in Figure 6-46, and apply textures to the objects in the scene, as shown in the same figure.

(Expected time: 30 min)

Figure 6-46 *Model of a study table after applying textures*

Chapter 7

Lighting

Learning Objectives

After completing this chapter, you will be able to:
- *Work with standard Maya lights*
- *Add glow and halo effects to lights*
- *Work with the camera*

INTRODUCTION

Lights are objects that produce real lighting effects like street lights, street lights, flash lights, house-hold lights and so on. When there is no light in a scene, the scene is rendered with default lighting. Moreover, light objects can be used to project the images. In this chapter, you will learn about various lights that you can use in your scene to give it realistic lighting effects.

TYPES OF LIGHTS

There are six types of lights in Maya. To create a light, choose **Create > Objects > Lights** from the menubar; a cascading menu will be displayed. Choose the required light from the cascading menu and click in the viewport; the light will be created in your scene. Different types of lights in Maya are discussed next.

Ambient Light

| **Menubar:** | Create > Objects > Lights > Ambient Light |

The ambient light is a single point light that projects the rays uniformly in all directions and lights up the scene. To create an ambient light, choose **Create > Objects > Lights > Ambient Light** from the menubar; ambient light will be created at the center of the viewport. You can modify the attributes of this light. To do so, select the ambient light in the viewport. Next, choose **Windows > Editors > General Editor > Attribute Editor** from the menubar; the **Attribute Editor** displaying the properties of the ambient light will be displayed on the right of the viewport. Some of the attributes in Maya are common for all lights.

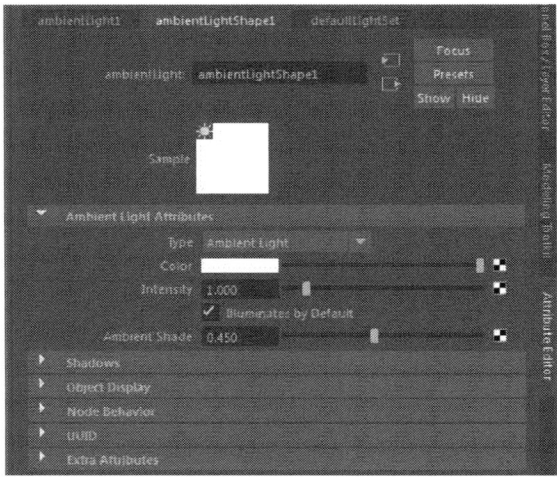

Figure 7-1 The **Ambient Light Attributes** area

if the value in this attribute is set to 1, the light will come from the point where the light is placed. In other words, the ambient light will act like a point light, when the value of the **Ambient Shade** attribute is 1. The default value of the **Ambient Shade** attribute is 0.45. Figure 7-2 shows the ambient light when the **Ambient Shade** value is set to **0.25** and Figure 7-3 shows the ambient light when the **Ambient shade** value is set to **1**.

Lighting

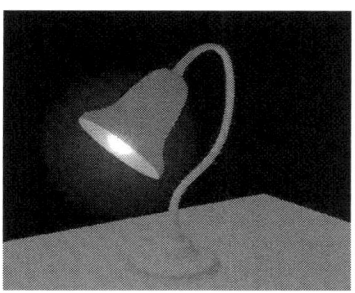 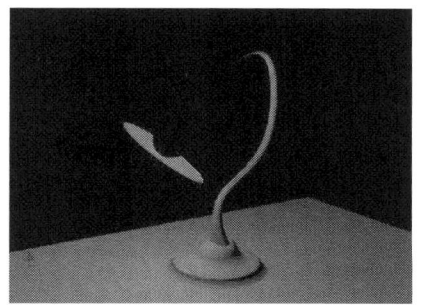

Figure 7-2 Ambient light with the **Ambient Shade** value set to **0.25**

Figure 7-3 Ambient light with the **Ambient Shade** value set to **1**

Directional Light

Menubar: Create > Objects > Lights > Directional Light

The directional light is used to create a distant point light. The light rays coming from the directional light are parallel to each other. To create a directional light, choose **Create > Objects > Lights > Directional Light** from the menubar; a directional light will be created on the grid in the viewport, as shown in Figure 7-4. You can also modify the attributes of this light. To do so, select the light in the viewport; the **Attribute Editor** showing the attributes of the directional light will be displayed on the right of the viewport. Some of the attributes of the directional light are same as those of the ambient light as discussed earlier.

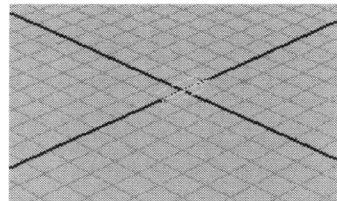

Figure 7-4 A directional light created

Point Light

Menubar: Create > Objects > Lights > Point Light

The point light is a single source of light which projects light evenly in all directions. To create a point light, choose **Create > Objects > Lights > Point Light** from the menubar; a point light will be created at the center of the viewport, as shown in Figure 7-5. Most of the attributes of the point light are similar to the attributes of the ambient light.

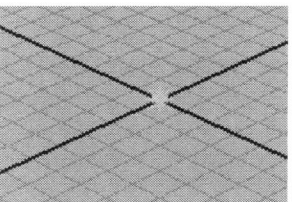

Figure 7-5 A point light created

Spot Light

Menubar: Create > Objects > Lights > Spot Light

The spot light evenly throws a beam of light within a narrow range in a conical shape, refer to Figure 7-6. Figure 7-7 shows a flower pot illuminated by a spot light. To create a spot light, choose **Create > Objects > Lights > Spot Light** from the menubar; the spot light will be created at the center of the viewport. Most of the attributes of the spot light are similar to those of the ambient light.

Figure 7-6 The spot light

Figure 7-7 A flower pot illuminated by a spot light

Area Light

Menubar: Create > Objects > Lights > Area Light

The area light is a type of light that has a two-dimensional rectangular light source. It emits light from a rectangular area. The larger the size of the light, the more illuminated the scene will be. To create an area light, choose **Create > Objects > Lights > Area Light** from the menubar; an area light will be created at the center of the viewport, as shown in Figure 7-8. It is used to create high quality still images. Therefore, the scene with an area light will take more time for rendering as compared to the other lights.

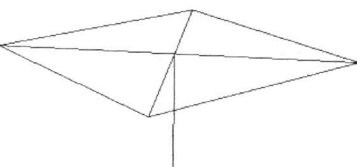

Figure 7-8 The area light

Volume Light

Menubar: Create > Objects > Lights > Volume Light

The volume light is used to add a volume light to the scene. This light is represented by the icon. The innermost area of the volume light icon represents the visual extent of the light. You can also use this light as a negative light. For example, you can use it to remove illumination from a particular area or to lighten up the dark shadows in the scene. To create a volume light, choose **Create > Objects > Lights > Volume Light** from the menubar; a volume light will be created at the center of the viewport, as shown in Figure 7-9. Figure 7-10 shows the effect of the volume light.

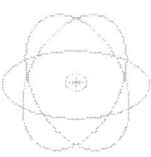

Figure 7-9 The volume light

Figure 7-10 The volume light effect

GLOW AND HALO EFFECTS

The glow and halo effects are used to add a realistic effect to the scene. These effects can be added to any light by using the **Attribute Editor**. To add these effects to a light, select the light in the viewport and choose **Windows > Editors > General Editors > Attribute Editor** from the menubar; the **Attribute Editor** displaying the attributes of the selected light will be displayed. Choose the checker box on the right of the **Color** attribute, refer to Figure 7-11; the **Create Render Node** window will be displayed in the viewport. Select the **Glow** option from the left pane of the **Create Render Node** window.

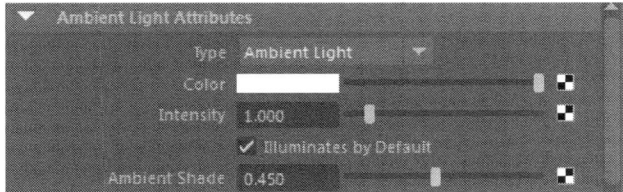

Figure 7-11 Choosing the checker box next to the **Color** attribute

Next, choose **Optical FX** from the right pane of the window; the **Optical FX Attributes** area will be displayed in the **Attribute Editor**, as shown in Figure 7-12.

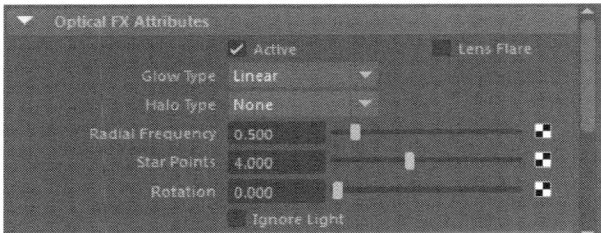

Figure 7-12 The **Optical FX Attributes** area

LIGHT LINKING

Light linking is a process of linking light to specific objects in a scene. To link lights, the light affects only the object to which it is linked in the scene. To link an object to the light, select the light and then select the **Rendering** menu from the **Menuset** drop-down list in the Status Line. Next, choose **Lighting/Shading > Light Linking > Light Linking Editor > Light-Centric** from the menubar; the **Relationship Editor** will be displayed in the viewport, refer to Figure 7-13. Select the light that you want to link from the **Light Sources** area and then select the objects from the **Illuminated Objects** area of the **Relationship Editor**; now the light source will illuminate only the linked objects. Next, close the **Relationship Editor**.

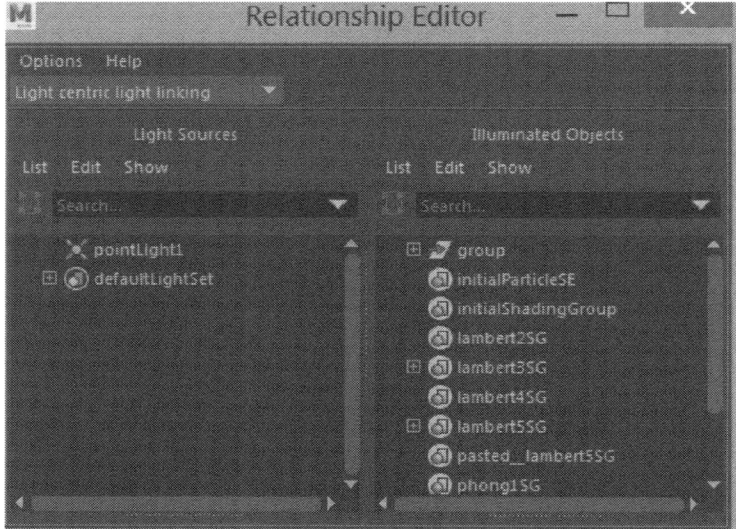

*Figure 7-13 The **Relationship Editor***

CAMERAS

In Maya, cameras are used to view a scene from different angles. These cameras work in a similar way as the still and video cameras in the real world. There are five types of cameras in Maya: Camera, Camera and Aim, Camera Aim and Up, Stereo Camera, and Multi-Stereo Rig. These camera types are created with the help of the **Camera**, **Camera and Aim**, **Camera Aim and Up**, **Stereo Camera**, and **Multi-Stereo Rig** tools, respectively.

Camera and Aim

Menubar: Create > Objects > Cameras > Camera and Aim

The **Camera and Aim** tool is used to create a basic camera and an aim vector control. To do so, choose **Create > Objects > Cameras > Camera and Aim** from the menubar; a camera will be created in the viewport, refer to Figure 7-14. This control is used to aim the camera at a specified point in the scene, refer to Figure 7-15.

Figure 7-14 A camera with an aim created

Figure 7-15 The camera aiming at a point in the scene

Lighting

Camera, Aim and Up

Menubar: Create > Objects > Cameras > Camera, Aim and Up

The **Camera, Aim and Up** tool is used to create a basic camera with an aim vector and an up vector controls. To do so, choose **Create > Objects > Cameras > Camera, Aim and Up** from the menubar; a camera will be created in the viewport, as shown in Figure 7-16.

This aim vector control is used to aim the camera at a specified point in the scene. The up vector control is used to rotate the camera, refer to Figure 7-17.

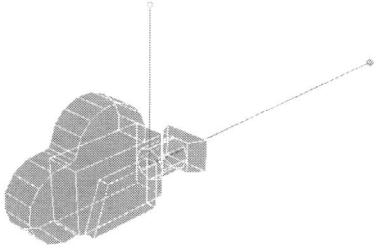 

Figure 7-16 A camera with the aim vector and up vector controls

Figure 7-17 A camera with the aim vector and up vector controls focusing on a point in the scene

Stereo Camera

Menubar: Create > > Objects > Cameras > Stereo Camera

The **Stereo Camera** tool is used to create stereoscopic cameras to produce an anaglyph or parallel image. This image when composited in a compositor produces renders with a depth illusion. To create a stereo camera, choose **Create > Objects > Cameras > Stereo Cameras** from the menubar; a stereo camera will be created in the viewport.

Multi Stereo Rig

Menubar: Create > Objects > Cameras > Multi Stereo Rig

The **Multi Stereo Rig** tool is used to create multi-camera rig for stereo cameras. By default, it is done by three layered camera rig.

TUTORIAL

All the files used in this tutorial can be downloaded from the CADSoft website (*www.cadsofttech.com*). These files are compressed in zip file format and are required to be extracted before using them in the tutorials. The path of the files is as follows: *Textbooks > Animation and Visual Effects > Maya > Autodesk Maya 2019 for Novices*

Tutorial 1

In this tutorial, you will create a scene in which the light is scattering through a cylindrical object, as shown in Figure 7-18. **(Expected time: 15 min)**

The following steps are required to complete this tutorial:

a. Create a project folder.
b. Create a cylinder.
c. Add a point light in the scene.
d. Save and render the scene.

Creating a Project Folder

Create a new project folder with the name *c07_tut1* at *\Documents\maya2019* and then save the file with the name *c07tut1* as discussed in Tutorial 1 of Chapter 2.

Creating a Cylinder

In this section, you will create a cylindrical object through which the light will scatter in all directions.

Figure 7-18 The light scattering through a cylinder

1. Maximize the top-Y viewport. Next, choose **Create > Objects > Polygon Primitives > Cylinder** from the menubar and then create a cylinder.

2. In the **Channel Box / Layer Editor**, expand the **polyCylinder1** node in the **INPUTS** area

Lighting

and set the parameters as follows:

Radius: **2** Height: **10**
Subdivisions Axis: **50** Subdivisions Height: **25**

3. In the **Channel Box / Layer Editor**, enter **2** in the **Scale X**, **Scale Y**, and **Scale Z** edit boxes and enter **0** in the **Translate X** and **Z** edit boxes, and **10** in the **Translate Y** edit box.

4. Maximize the persp viewport. Next, press and hold the right mouse button on the cylinder; a marking menu is displayed. Choose **Face** from the marking menu; the face selection mode is activated. Next, delete the top and bottom faces of the cylinder.

5. Delete some more faces of the cylinder randomly, as shown in Figure 7-19.

6. Choose **Create > Objects > Polygon Primitives > Plane** from the menubar. Next, create a plane in the persp viewport. Now, choose **Move Tool** from the Tool Box and place the plane below the cylinder.

7. In the **Channel Box / Layer Editor**, expand the **polyPlane1** node in the **INPUTS** area and set the parameters as follows:

Width: **80** Height: **80**

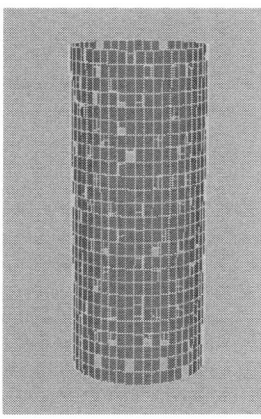

Figure 7-19 Randomly deleted faces

8. Make sure the **Modeling** menuset is selected from the **Menuset** drop-down list in the Status Line. Next, press and hold the right mouse button on the cylinder; a marking menu is displayed. Choose **Face** from the marking menu; the face selection mode is activated. Select all the faces of the cylinder. Next, choose **Edit Mesh > Components > Extrude** from the menubar; the **polyExtrudeFace1** In-View Editor is displayed in the viewport. Now, enter **0.6** in the **Thickness** edit box of the **polyExtrudeFace1** In-View Editor. Press W to exit the tool.

9. Press and hold the right mouse button on the cylinder; a marking menu is displayed. Choose **Object Mode** from the marking menu; the object selection mode is activated.

Adding a Point Light to the Scene

In this section, you will add a point light to the scene. It will act as the source of light.

1. Choose **Create > Objects > Lights > Point Light** from the menubar; a point light is created in the viewport.

2. In the **Channel Box / Layer Editor**, enter **10** in the **Translate Y** edit box.

3. Make sure the point light is selected in the viewport. Next, press CTRL+A; the **pointLightShape1** tab is displayed in the **Attribute Editor** with the attributes of the point light.
4. In the **Point Light Attributes** area, click on the color swatch on the right of the **Color** attribute; the **Color History** palette is displayed. Set the color of the light to orange. Alternatively, select **RGB, 0 to 1.0** option from the drop-down list located at the bottom right of the **Color History** palette and then set the values as given below:

   R: **0.667**   G: **0.35**   B: **0.078**

   Next, click anywhere outside the palette to close it and then enter **30** in the **Intensity** edit box.

5. In the **Light Effects** area of the **pointLightShape1** tab, choose the checker button on the right of the **Light Fog** attribute; the **lightFog1** tab is displayed in the **Attribute Editor**.

6. Select the point light in the viewport. Make sure that the **pointLightShape1** tab is chosen, in the **Attribute Editor** and then set the parameters in the **Light Effects** area as follows:

   Fog Radius: **40**   Fog Intensity: **4**

7. Choose the checker button on the right side of the **Light Glow** attribute; the **opticalFX1** tab is displayed in the **Attribute Editor**.

8. Select the point light in the viewport. In the **pointLightShape1** tab in the **Attribute Editor**, expand the **Shadows** area and select the **Use Depth Map Shadows** check box in the **Depth Map Shadow Attributes** area.

9. In the **Depth Map Shadow Attributes** area, set the parameters as follows:

   Resolution: **1024**   Fog Shadow Samples: **50**

10. In the **sphereShape#** tab of the **Attribute Editor**, expand the **Render Stats** area. Next, select the **Volume Samples Override** check box and enter **2** in the **Volume Samples** edit box.

Saving and Rendering the Scene

In this section, you will save the scene that you have created and then render it. You can

Lighting

view the final rendered image of the scene by downloading the *c07_maya_2019_rndr.zip* file from *www.cadsofttech.com*. The path of the file is as follows: *Textbooks > Animation and Visual Effects > Maya > Autodesk Maya 2019 for Novices*.

1. Choose **File > Save Scene** from the menubar.

2. Maximize the persp viewport if not already maximized. Choose the **Render the current frame** button from the Status Line; the **Render View** window is displayed. This window shows the final output of the scene, refer to Figure 7-18.

EXERCISE

The rendered output of the models used in the following exercise can be accessed by downloading the *c07_maya_2019_exr.zip* file from *www.cadsofttech.com*. The path of the file is as follows: *Textbooks > Animation and Visual Effects > Maya > Autodesk Maya 2019 for Novices*.

Exercise 1

Create and texture the scene shown in Figure 7-20 and then add lights to it to get the output shown in Figure 7-21. **(Expected time: 15 min)**

Figure 7-20 *The textured scene*

Figure 7-21 *The physical sun and sky effect applied to the scene*

This page is intentionally left blank

Chapter 8

Animation

Learning Objectives
After completing this chapter, you will be able to:
- *Understand the basic concepts of animation*
- *Understand different types of animation*
- *Use the Graph Editor for editing animation*
- *Use Animation layers*

INTRODUCTION

Animation is a process of displaying a sequence of images in order to create an illusion of movement. In this chapter, you will animate models using various animation techniques such as keyframe animation, path animation, nonlinear animation, and technical animation.

To animate a 3D object, you need to record its position, rotation, and scale on different frames. These frames are known as keyframes. The keys between the keyframes contain information about the actions performed in the animation. When an animation is played, the frames are displayed one after the other in quick succession which creates an optical illusion of motion. In this chapter, you will also learn about the playback control buttons available at the bottom of the interface and various additional tools used for creating an animation.

ANIMATION TYPES

In Maya, you can create animation using different techniques such as Keyframe Animation, Effects Animation, Nonlinear Animation, Path Animation, Motion Capture Animation, and Technical Animation. Some of these techniques are discussed below.

Keyframe Animation

The Keyframe animation is used to animate objects by manually setting the keyframes over time. It is the most commonly used animation type as it is highly flexible and helps to create complex animations easily.

Effects Animation

The Effects animation is also known as the dynamic animation. It is used to create and simulate physical phenomena such as fire and smoke. Animation of fluids, particles, and hair/fur are some examples of effects animation.

Nonlinear Animation

The Nonlinear animation is an advanced method of animation. It is used to blend, duplicate, and split animation clips to achieve different motion effects. The nonlinear animation is controlled by using the **Trax Editor**. For example, you can loop the walk cycle of your character by using Graph Editor.

Path Animation

The Path animation is used to animate an object's translation and rotation attributes on the basis of a NURBS curve. This type of animation is used to animate an object along a path such as a moving car on the road or a moving train on the railtrack.

Motion Capture Animation

The Motion Capture animation is the process of recording human body movement for immediate or delayed analysis and playback. It is used to animate a character by using the motion capturing devices. A motion capture device helps in real time monitoring and recording of data.

Technical Animation

The Technical Animation is used to animate an object by linking the translation and rotation attributes of one object with another object. The linking is done by setting driven keys in such a way that the attributes of one object are governed by the attributes of another object. For example, if you want to animate a locomotive engine, you need to link various parts of the engine by using the technical animation.

ANIMATION CONTROLS

In Maya, you can edit and view an animation. It can be done using various buttons such as Playback Controls, Animation Preferences, and so on. Some of these buttons are discussed below.

Playback Controls

The Playback Controls, shown in Figure 8-1, are used to control the animation in a scene. These buttons are located on the Time Slider at the bottom of the interface.

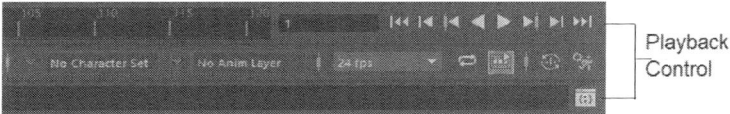

Figure 8-1 The playback controls

Animation preferences

The **Animation preferences** button is used to display the **Preferences** window with the **Time Slider** option selected by default in the **Categories** list. The options in this window are used to edit the animation settings in Maya. To do so, choose the **Animation preferences** button located below the **Go to end of playback range** button in the playback control area; the **Preferences** window will be displayed in the viewport, as shown in Figure 8-2.

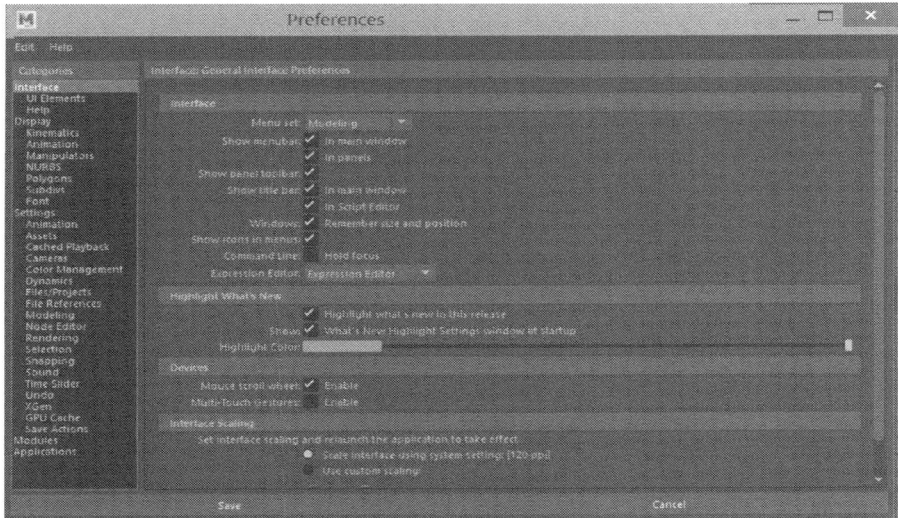

Figure 8-2 The **Preferences** window

COMMONLY USED TERMS IN ANIMATION

In Maya, some terms are used very commonly. These terms are discussed next.

Frame Rate

The frame rate is termed as the number of frames or images displayed per second in a sequence. It is abbreviated as fps (frames per second). It is the total number of frames played per second in an animation.

Range

The term range is used to define the total length of an animation. The range of an animation is calculated in frames. For calculating the range of an animation, multiply the frame rate with the total time of animation. For example, if you have a frame rate of 24 fps and the total time of the animation is 5 secs, then the range of the animation will be: 24 X 5 = 120 frames.

Setting Keys

Setting keys is defined as the process of specifying the translational, rotational, and scale values of an object on a particular frame. For example, to set a key for translation of an object, select the object that you need to animate and choose a frame in the timeline on which you want to set the key. Then, select the **Animation** menuset from the **Menuset** drop-down list in the Status Line. Next, choose **Key > Set > Set Key** from the menubar; the key will be set at the selected frame in the timeline. In the **Channel Box / Layer Editor**, press and hold the right mouse button on any translate axis; a flyout will be displayed. Choose **Key Selected** from the flyout; the key for the selected translate axis will set. On setting the keys, the default background color of the attributes in the **Channel Box / Layer Editor** will change to peach color, indicating that the keys are set for the selected attributes.

Tip
You can set the keys for animation by pressing the S key.

UNDERSTANDING DIFFERENT TYPES OF ANIMATIONS

In the beginning of this chapter, you learned in brief about different types of animations. Now, you will learn to animate objects using some of these animation types.

Path Animation

The path animation method is used to animate an object along a path. To do so, activate the top-Y viewport, choose **Create > Objects > Curve Tools > EP Curve Tool** from the menubar, and then create a curve, refer to Figure 8-3. Next, choose **Create > Objects > NURBS Primitives > Sphere** from the menubar and create a sphere in the viewport, as shown in Figure 8-3. Now, press and hold the SHIFT key and select the sphere first and then the curve. Next, select the **Animation** menuset from the **Menuset** drop-down list in the Status Line. Next, choose **Constrain > Motion Paths > Attach to Motion Path** from the menubar; the sphere will be attached to the curve. Choose the

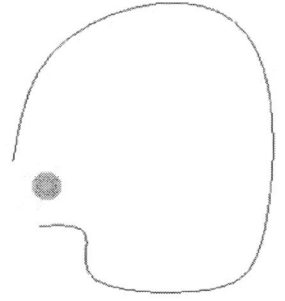
Figure 8-3 The NURBS curve and the sphere

Play forwards button from the playback controls; the sphere will start moving along the path. You can also use a closed path to animate an object.

Sometimes, when you choose the **Play forwards** button from the animation playback controls, the sphere may not sail smoothly on the curve. To overcome this problem, select the sphere from the viewport and choose **Constrain > Motion Paths > Flow Path Object** from the menubar; a lattice will be created for the object throughout the curve. The lattice provides smoothness to the motion of the sphere. To detach the sphere from the curve, select it. In the **Channel Box / Layer Editor**, press and hold the SHIFT key and select the **Translate X**, **Translate Y**, **Translate Z**, **Rotate X**, **Rotate Y**, and **Rotate Z** options. Now, press and hold the right mouse button over the selected attributes; a flyout will be displayed. Choose **Break Connections** from the flyout; the sphere will be detached from the curve. Similarly, you can detach the curve from the sphere.

Tip
To animate multiple objects on a single path curve, select the objects that you want to animate and then choose the path curve; all the objects will get attached to the path through their pivot points.

Note
*If object gets distorted on applying the **Flow Path Object** option, choose **Windows > Editors > Outliner** from the menubar and select the **FFD1 lattice** and **FFD1Base** options from the **Outliner** window. Then, scale the two selected lattices such that the object fits well into the lattice structure.*

Keyframe Animation

The keyframe animation method is the standard method used for animating an object. It is used to animate an object by creating smooth transitions between different keyframes. This is done by setting the keys for the object at two extreme positions. Maya interpolates the value for the keyed attributes with the change in the timeline between the two set keys.

You can set the key for animating an object by pressing S on the keyboard. Alternatively, select the frame at which you want to set the key and choose **Key > Set > Set Key** from the menubar; a keyframe will be set at the selected frame. You can also use the auto keyframe method to set the keys for creating an animation. To do so, activate the persp viewport and choose **Create > Objects > Polygon Primitives > Cube** from the menubar; to create a polygon cube in the viewport. Next, choose the **Animation preferences** button located below the animation playback controls; the **Preferences** window will be displayed with the **Time Slider: Animation Time Slider and Playback Preferences** area on the right of the **Preferences** window, as shown in Figure 8-4. Set the required parameters in this window and choose the **Save** button.

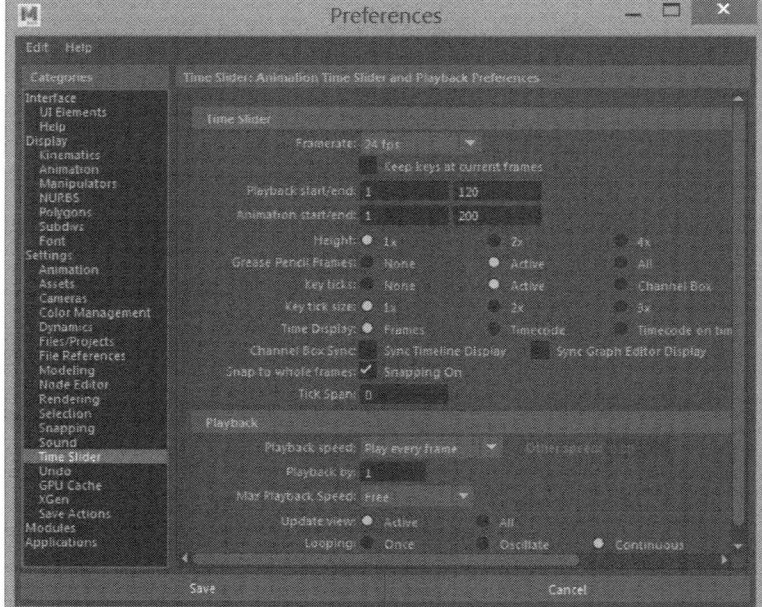

*Figure 8-4 The **Time Slider: Animation Time Slider and Playback Preferences** area in the Preferences window*

Select the polygon cube from the viewport and choose **Modify > Transform > Freeze Transformations** from the menubar; the move, rotate and scale attributes of the cube will be set to **0**. Set the Time Slider to frame **1**. Next, choose **Key > Set > Set Key** from the menubar; the key will be set on frame **1** and then set the value in the Time Slider to frame **300**. Next, set the **Translate X** value in the **Channel Box / Layer Editor** to **15** and choose **Key > Set > Set Key** from the menubar; the key will be set for the selected frame. Now, choose the **Play forwards** button from the playback controls to preview the animation. Move the Time Slider to frame **150**. Next, set the **Translate Z** value to **5**. Now, choose **Key > Set > Set Key** from the menubar to set the key. Then, choose the **Play forwards** button from the playback controls to preview the animation; the cube will move along the curve path.

If you want the cube to rotate while animating, select the cube and set the current time indicator to frame **0** on the timeline. Now, press SHIFT+E to set the keys for rotation. In the **Channel Box / Layer Editor**, set the Time Slider to frame **24** and set the **Rotate Y** attribute to **300** in the **Channel Box / Layer Editor**. Then, press SHIFT+E again; the keys for rotation will be set. Finally, choose the **Play forwards** button from the animation playback controls to preview the animation; the cube will rotate on its own axis. To translate the position of the cube in the viewport, move the current time indicator to frame **1** in the timeline. Now, press SHIFT+W to set the keys for translation. Next, move the current time indicator to frame **24** and set the **Translate X** attribute to **20** in the **Channel Box / Layer Editor**. Press SHIFT+W to set the translate key and choose the **Play forwards** button from the playback controls to preview the animation; the cube will translate and rotate simultaneously. You can also set the keys for animation by enabling the **Auto keyframe toggle** button in the timeline. It is a toggle button which turns red when active and sets the keys automatically, refer to Figure 8-5.

Nonlinear Animation

The nonlinear animation is used to animate an object that is independent of time. These clips can be used repeatedly to add motion to your scene which saves a lot of time for creating the animation. To apply nonlinear animation to an object, you need to use the **Trax Editor**. To display the **Trax Editor**, choose **Windows > Editors > Animation Editors > Trax Editor** from the menubar; **Trax Editor** will be displayed, as shown in Figure 8-6.

Figure 8-5 The *Auto keyframe toggle* button

Figure 8-6 The *Trax Editor*

The Playback Menu

The **Playback** menu consists of various options that are used to modify the animations as required. The **Playback** options is discussed next.

Playblast

The **Playblast** option is used to create a low resolution preview of the animation that you can use to review the animation and rectify errors. You can also change the format and quality of the playblast by using this option. To do so, choose **Playback > Playblast > Option Box** from the menubar; the **Playblast Options** window will be displayed, refer to Figure 8-7. You can set the options in the window as required.

Cached Playback

Cached playback is the process that continuously evaluate the animation and helps to speed up the animation playback in the viewport. This appears as a blue stripe running along the bottom of the Time Slider. The **Cached Playback** button is used to change the animation without the need to create a playblast.

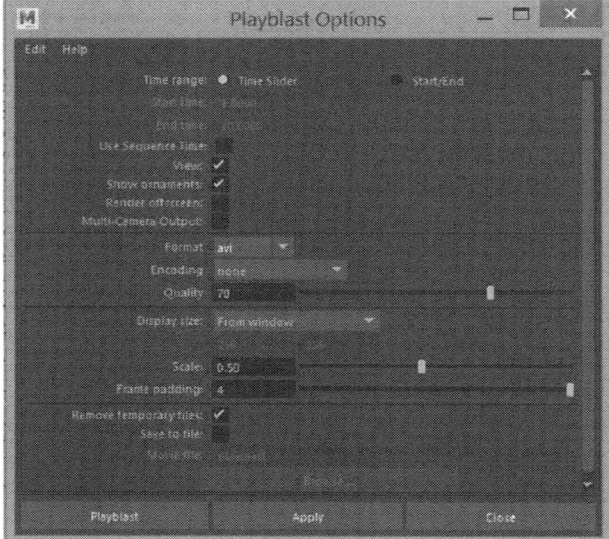

*Figure 8-7 The **Playblast Options** window*

Graph Editor

Menu: Windows > Editors > Animation Editors > Graph Editor

The **Graph Editor** is used to edit the animation curves. This window displays graphical representation of the animated object in the viewport. The graph helps you to change or set the values of keys in this window as required. The **Graph Editor** is used to store all the information about animation and provides you a direct access to fine-tune the animation. Each animation in Maya generates a value vs time graph. In this graph, the horizontal axis represents the time and the vertical axis represents the value. In the **Graph Editor**, the keyframes are represented by points on curves. You can move these points freely to fine-tune the animation. To move a point on the curve, select a key, press and hold the middle mouse button, and then drag the point in the timeline to adjust the animation as required. You can also snap the keys to the grids in the editor using the snap icons from the **Graph Editor** toolbar.

Move Nearest Picked Key Tool

The **Move Nearest Picked Key Tool** works on a single key at a time. It is different from the **Move Tool** as it moves keys individually. Select a key from the timeline and then choose the **Move Nearest Picked Key Tool** button from the toolbar. Now, press the middle mouse button in the **Graph Editor** to move the selected key for making changes in the animation.

Insert Keys Tool

The **Insert Keys Tool** is used to add a new key to an animation curve. To do so, select the curve on which you want to add a new key and then choose the **Insert Keys Tool** button from the **Graph Editor** toolbar. Now, click the middle mouse button on the selected curve; a new key will be created without changing the shape of the original animation curve.

Animation

Lattice Deform Keys
The **Lattice Deform Keys** tool is used to draw a lattice around a group of keys in the **Graph Editor** so that the selected keys can be transformed uniformly. To transform the keys using this tool, choose the **Lattice Deform Keys Tool** button from the **Graph Editor** toolbar. Next, press and hold the left mouse button and then select the keys in the **Graph Editor**; a lattice will be formed around the selected keys. Now, you can deform the lattice to transform the selected keys. This tool provides a high level of control over animation.

Region Tool: Scale or move keys
The **Region Tool: Scale or move keys** tool is used to move or scale the selected keys in the **Graph Editor**. To do so, choose the **Region Tool: Scale or move keys** button from the **Graph Editor** toolbar. Next, select the key on the curve and then move the key in any direction by using the middle mouse button.

Retime Tool: Scale and ripple keys
The **Retime Tool: Scale and ripple keys** tool is used to directly adjust the timing of key movements in an animation sequence. To adjust the timing, choose the **Retime Tool: Scale and ripple keys** tool from the **Graph Editor** toolbar. Next, double-click in the graph to create retime markers around segments of the animation curves you want to adjust.

Auto tangents
The **Auto tangents** tool is used to make the selected curve smooth by automatically adjusting the keys on the curve. By default, this tangent type is turned off.

Spline tangents
The **Spline tangents** tool is used to adjust the tangents on a curve so that curve becomes smoother. To adjust the tangents, select an animation key on the animation curve in the **Graph Editor** and then choose the **Spline tangents** tool from the **Graph Editor** toolbar. Alternatively, choose **Tangents > Spline** from the **Graph Editor** menubar, as shown in Figure 8-8.

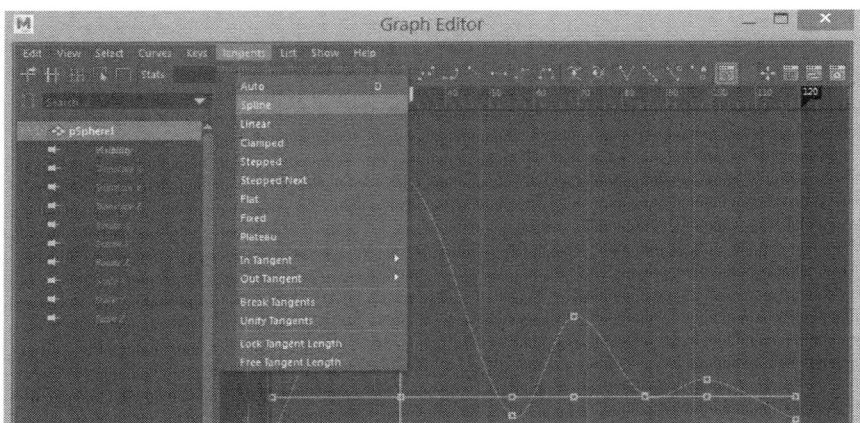

*Figure 8-8 The **Tangents** menu in the **Graph Editor***

Clamped tangents

The **Clamped tangents** tool has the characteristics of both the **Spline tangents** and the **Linear tangents** tools and it works similar to these tools.

Linear tangents

The **Linear tangents** tool is used to create a straight animation curve by joining two keys on the selected curve. Figures 8-9 and 8-10 show the animation curve before and after using the **Linear tangents** tool.

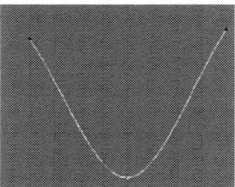 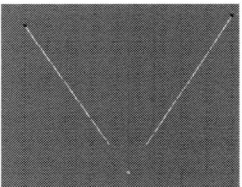

Figure 8-9 The animation curve before using the **Linear tangents** tool

Figure 8-10 The animation curve after using the **Linear tangents** tool

Note
The process of using or accessing the remaining tangent tools is similar as discussed for the **Spline tangents** *tool.*

Flat tangents

The **Flat tangents** tool is used to set the tangent of the selected curves horizontally. When you throw a ball up in the air, the ball stays at the topmost point for a moment before it comes down. To represent such an animation, you can use the **Flat tangents** tool. Figures 8-11 and 8-12 show the animation curve before and after using the **Flat tangents** tool.

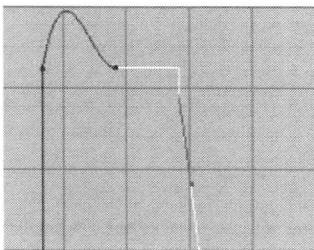 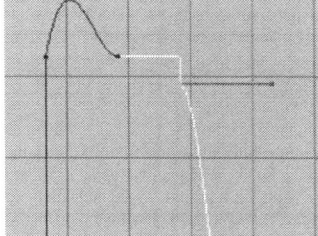

Figure 8-11 The animation curve before using the **Flat tangents** tool

Figure 8-12 The animation curve after using the **Flat tangents** tool

Step tangents

The **Step tangents** tool is used to change a flat curve in the shape of steps, refer to

Figures 8-13 and 8-14. You can also create the effect of the blinking light using this tool.

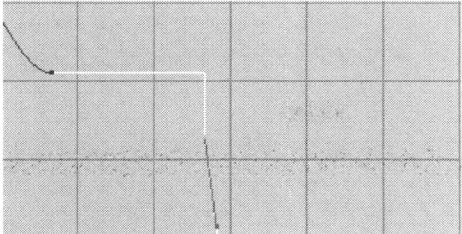 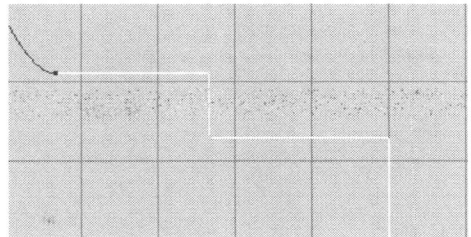

Figure 8-13 The animation curve before using the **Step tangents** tool

Figure 8-14 The animation curve after using the **Step tangents** tool

Plateau tangents

The **Plateau tangents** tool works similar to the **Spline tangents** and **Clamped tangents** tools. It is used to set the animation curves in such a way that they do not go beyond the position of their respective keyframes, refer to Figures 8-15 and 8-16.

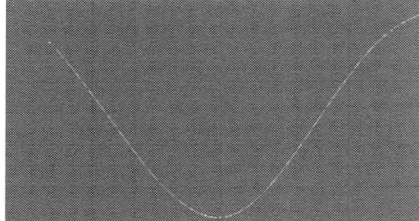 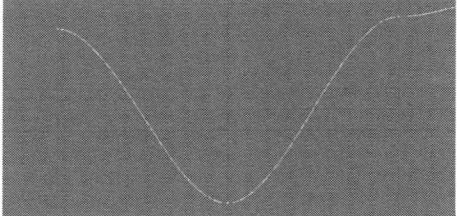

Figure 8-15 The animation curve before using the **Plateau tangents** tool

Figure 8-16 The animation curve after using the **Plateau tangents** tool

Buffer curve snapshot

The **Buffer curve snapshot** tool is used to take a snapshot of the selected curve. To take a snapshot, select the curve. Next, invoke the **Buffer Curve Snapshot** tool from the **Graph Editor** toolbar; the buffer curve snapshot will be taken for the selected curve. To view the buffer curve snapshot, choose **View > Show Buffer Curves** from the **Graph Editor** menubar, as shown in Figure 8-17.

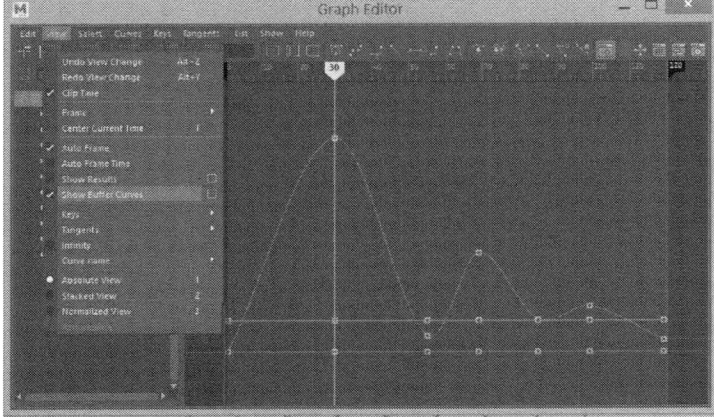

Figure 8-17 Choosing the **Show Buffer Curves** option from the **View** menu in the **Graph Editor** menubar

Tip
Tangents can be edited using the marking menus displayed by pressing CTRL +SHIFT and the middle mouse button.

Swap buffer curve

The **Swap buffer curve** tool is used to swap between the original curve and the edited curve. You can use the **Buffer curve snapshot** tool and the **Swap buffer curves** tool to compare the changes made in the animation curve. The changes in the animation curve will be indicated by a grey line.

Break tangents

The **Break tangents** tool is used to break the tangents joined to a key such that both handles of the broken tangent work separately to fine-tune the animation. Note that the broken tangent will be displayed in blue color.

Unify tangents

The **Unify tangents** tool is used to retain tangents at their original location. This tool works in such a way that if you manipulate changes in one tangent, the other tangent of the key will be equally affected. If you break two tangents, which are joined to a key using the **Break tangents** tool and then apply the **Unify tangents** tool on them, the two tangents will start acting as a single tangent.

Value snap on/off

The **Value snap on/off** tool is used to move the keys in the graph view to their nearest integer value by applying force on them.

Enable normalized curve display

The **Enable normalized curve display** tool is used to activate the normalized view. In this mode, the large key values are scaled down or small key values are scaled up to fit within -1 to 1 range. You can press the 3 key to activate the absolute view.

Disable normalized curve display

The **Disable normalized curve display** tool is used to disable the normalized view. You can press the 3 key to deactivate the absolute view.

Renormalize curves

The **Renormalize curves** tool in the **Graph Editor** toolbar is used to quickly normalize the selected curve to fit the key values of the selected animation curves within the range of normalization. The normalization range is between -1 and 1.

Enable stacked curve display

The **Enable stacked curve display** tool is used to display individual curves in stack. In this stacked view mode, no overlapping of curves is displayed. Each curve displays its own value axis which is normalized between 1 and -1, by default. You can press the 2 key to activate the absolute view.

Disable stacked curve display
The **Disable stacked curve display** tool is used to all curve overlapping. In this stacked view mode, you can't modify the time value.

Pre-infinity cycle
The **Pre-infinity cycle** tool is used to copy a selected animation curve and then repeat the animation infinitely in the graph view before the selected curve. The copied animation curve will be displayed as a dotted line, as shown in Figure 8-18.

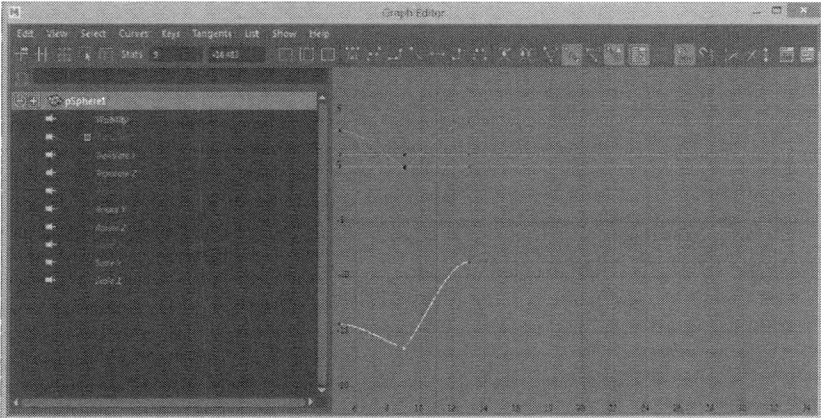

Figure 8-18 The pre-infinity cycle graph in the **Graph Editor**

Pre-infinity cycle with offset
The **Pre-infinity cycle with offset** tool is also used to repeat the selected animation curve infinitely through the graph view. This tool differs from the **Pre-infinity cycle** tool as it adds the first key value of the original curve to the last key value of the cycled curve.

Post-infinity cycle
The **Post-infinity cycle** tool is used to copy an animation curve and then join it after the same curve infinite number of times. Therefore, unlike the **Pre-infinity cycle** tool, this tool copies the animation curve and repeats it after the curve. The copied animation curve will be displayed as a dotted line.

Post-infinity cycle with offset
The **Post-infinity cycle with offset** tool is used to cycle curve with offset after its first key. It works similar to the **Pre-Infinity cycle with offset** tool, except that on using this tool the last key value of the original curve is added to the first key value of the cycled curve.

Unconstrained drag
The **Unconstrained drag** tool is used to constrain the movement of drag of the selected curve in the X and Y directions. To do so, press the left mouse button on the **Unconstrained drag** tool; the tool icon will change to **Constrained x-axis drag**. Now, select the tool and then press the middle mouse button in the **Graph Editor** to move the selected curve in the x-axis only.

Again, press the left mouse button on the **Unconstrained drag** tool; the tool icon will change to **Constrained y-axis drag**. Press the middle mouse button in the **Graph Editor** to move the selected curve in the y-axis only.

Open the Dope Sheet

The **Open the Dope Sheet** tool is used to switch between the **Graph Editor** and the **Dope Sheet** to set the animation keys of the current object into the **Dope Sheet** area, refer to Figure 8-19. The **Dope Sheet** window is used to display the time horizontally in blocks. To invoke the **Dope Sheet**, choose **Windows > Editors > Animation Editors > Dope Sheet** from the menubar.

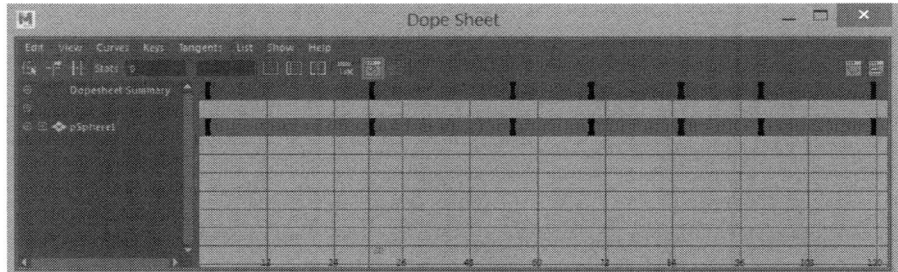

Figure 8-19 The Dope Sheet

Open the Trax Editor

The **Open the Trax Editor** tool is used to load the **Trax Editor** along with the animation clips of the current object. To load it, choose **Windows > Editors > Animation Editors > Trax Editor** from the menubar; the **Trax Editor** window will be displayed. In this editor, you can position, scale, cycle, and blend the animation sequences as required.

ANIMATION LAYERS

Animation layers are used to add or blend two animations together. In other words, these layers help you to organize a keyframe animation without overlapping the original animation. You can control these animations using the **Animation Layer Editor**. To open the animation layer, select the object; the **Channel Box / Layer Editor** will be displayed on the right of the viewport. To activate the **Animation Layer Editor**, choose the **Anim** tab from the **Channel Box / Layer Editor**; the attributes for the animation will be displayed in the **Channel Box / Layer Editor**, as shown in Figure 8-20. To set the **Animation Layer Editor** as a floating window, choose **Show > Floating Window** from the **Channel Box / Layer Editor**; the Animation Layer Editor floating window will be displayed. You can create a number of animation layers using this editor.

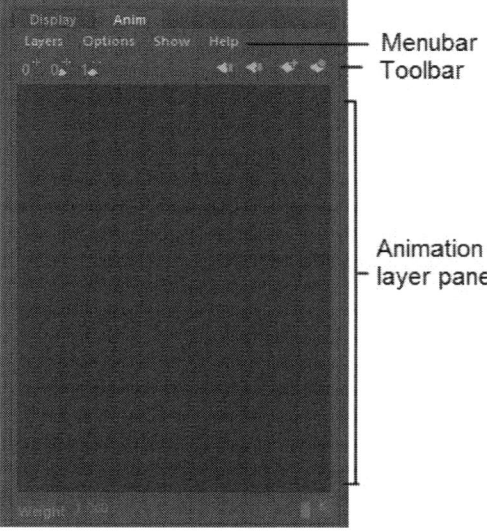

Figure 8-20 The Animation Layer Editor

Animation

Animation Layer Pane

The animation layer pane displays the hierarchy of animation layers that have been created. By default, the animation layers in this pane are arranged from bottom to top. Whenever you create a new layer, it gets added at the top of the **Animation Layer Pane**. You can change the arrangement of these layers by choosing **Options > Reverse Layer Stack** from the **Animation Layer Editor** menubar. On doing so, the layers will be arranged from top to bottom. Also, all newly created layers will be added at the bottom of the layer stack.

Creating the Parent-Child Relationship in the Animation Layer Editor

The Animation Layer **hierarchy** is used to parent and unparent an animation layer. To create a parent-child relationship between layers, select a layer from the Animation Layer Editor, drag it using the middle mouse button and drop it over another layer. The layer on which another layer is dropped will now act as the parent layer of the dropped layer. Also, a down arrow will be displayed in the parent layer, as shown in Figure 8-21.

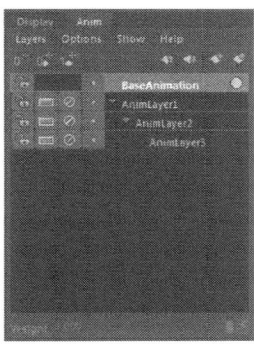

Figure 8-21 Layers showing the parent-child relationship

Similarly, you can create any number of parent-child relationships in the Animation Layer Editor. You can also unparent a layer in the **Animation Layer Editor**.

TUTORIALS

All the files used in the tutorials can be downloaded from the CADSoft website (*www.cadsofttech.com*). These files are compressed in zip file format and are required to be extracted before using them in the tutorials. The path of the files is as follows: *Textbooks > Animation and Visual Effects > Maya > Autodesk Maya 2019 for Novices*

Tutorial 1

In this tutorial, you will create a bouncing ball animation. **(Expected time: 30 min)**

The following steps are required to complete this tutorial:

a. Create the project folder.
b. Create a ball.
c. Create and refine the animation.
d. Save and render the scene.

Creating the Project Folder

Create a new project folder with the name *c08_tut1* at *\Documents\maya2019* and then save the file with the name *c08tut1*, as discussed in Tutorial 1 of Chapter 2.

Creating the Model of a Ball

In this section, you will create a ball using the polygon sphere.

1. Activate the top-Y viewport. Choose **Create > Objects > Polygon Primitives > Sphere** from the menubar and create a sphere in the viewport.

2. In the **Channel Box / Layer Editor**, expand the **polySphere1** node in the **INPUTS** area and make sure **1** is entered in the **Radius** edit box.

3. In the **Channel Box / Layer Editor**, click on **pSphere1**; a text box is activated. Next, enter **ball** in the text box and press ENTER; the **pSphere1** is renamed as *ball*.

4. Choose **Create > Objects > Polygon Primitives > Plane** from the menubar and create a plane below the *ball*. Next, activate the persp viewport. Choose **Move Tool** from the Tool Box and place *ball* on the plane, as shown in Figure 8-22.

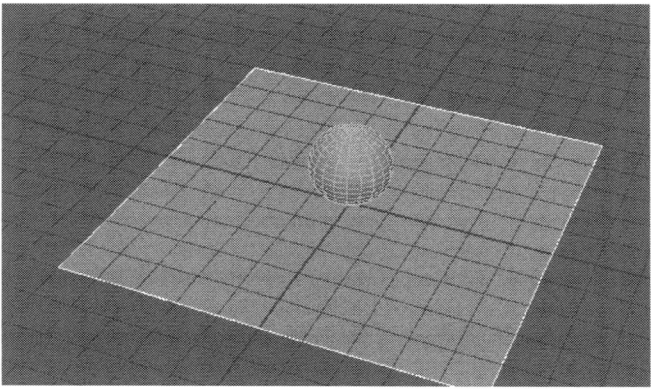

Figure 8-22 Ball placed on the plane

5. Select the plane in the persp viewport. Expand the **polyPlane1** area in the **INPUTS** node of **Channel Box / Layer Editor** and set the parameters as follows:

Width: **30**   Height: **30**

Create and Refine the Animation

In this section, you will set the animation keys to create the bouncing ball animation.

1. Select the **Animation** menuset from the **Menuset** drop-down list.

2. Select *ball* in the viewport and then choose **Modify > Transform > Freeze Transformations** from the menubar; the transformation values of *ball* are set to 0.

3. Choose **Windows > Editors > Settings/Preferences > Preferences** from the menubar; the **Preferences** window is displayed. Choose **Time Slider** from the **Categories** list of the window; the **Time Slider: Animation Time Slider and Playback Preferences** area is displayed in the **Preferences** window.

4. Enter **50** in the second edit box corresponding to the **Playback start/end** attribute. Next, make sure the **24fps x 1** option is selected from the **Playback speed** drop-down list in the **Playback** area, refer to Figure 8-23. Next, choose the **Save** button; the **Preferences** window closes.

Animation

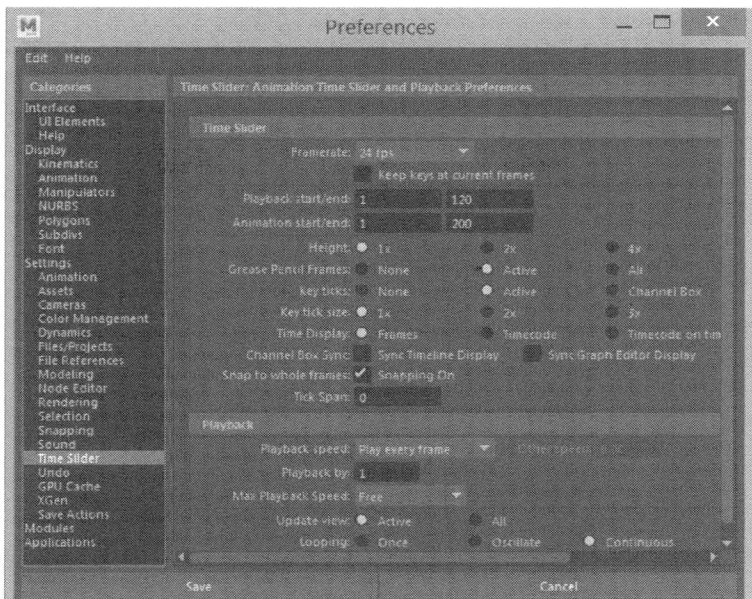

*Figure 8-23 Setting the options in the **Preferences** window*

5. Make sure *ball* is selected and the frame 1 is selected in the timeline. In the **Channel Box / Layer Editor**, enter **9** in the **Translate Y** edit box; *ball* moves upward along the Y axis, refer to Figure 8-24. Next, choose **Key > Set > Set Key**; the key is set at frame 1. Alternatively, press the S key to set the key.

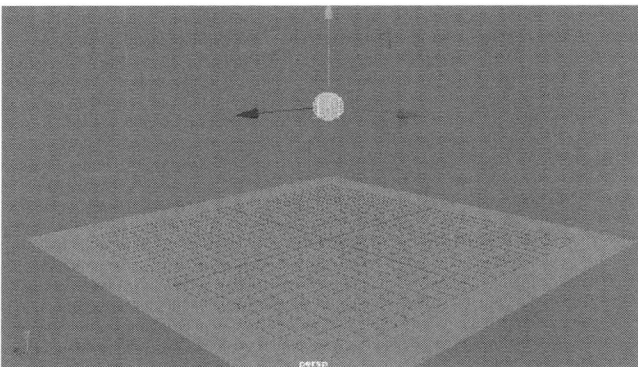

Figure 8-24 Position of the ball changed

6. Select frame 25 and enter **-0.002** in the **Translate Y** edit box of the **Channel Box / Layer Editor**; *ball* touches the plane. Next, press the S key; the key is set at frame 25.

7. Select frame 50 and then enter **9** in the **Translate Y** edit box of the **Channel Box / Layer Editor**; *ball* moves upward along the Y axis. Next, press the S key; the key is set at frame 50.

Next, you will scale down *ball* at frame 25.

8. Select frame 25 and enter **0.85** in the **Scale Y** edit box of the **Channel Box / Layer Editor**; *ball* gets squashed, as shown in Figure 8-25. Also, move the *ball* downward along the Y axis such that it touches the plane. Next, press the S key; the key is set at frame 25.

9. Select frame 24 and enter **1** in the **Scale Y** edit box of the **Channel Box / Layer Editor**; *ball* gets stretched. Now, enter **0** in the **Translate Y** edit box of the **Channel Box / Layer Editor**; *ball* moves upward. Next, press the S key; the key is set at frame 24.

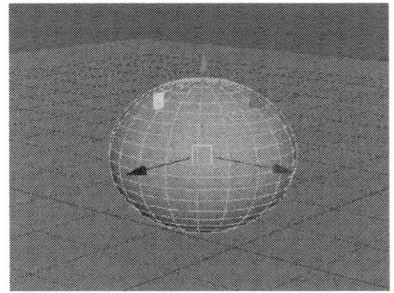

Figure 8-25 The ball squashes

10. Select frame 26 and enter **1** in the **Scale Y** edit box of the **Channel Box / Layer Editor**; *ball* gets stretched. Now, enter **0** in the **Translate Y** edit box of the **Channel Box / Layer Editor**; *ball* moves upward. Next, press the S key; the key is set at frame 26.

11. Choose the **Play forwards** button from the playback control area to preview the animation; *ball* starts bouncing like a rubber ball.

Note
*If the ball penetrates into the plane, you need to adjust the **Translate Y** value at frames 25 and 26.*

Saving and Rendering the Scene

In this section, you will save the scene that you have created and then render it. You can view the final rendered image sequence of the scene by downloading the *c08_maya_2019_rndr.zip* file from *www.cadsofttech.com*. The path of the file is as follows: *Textbooks > Animation and Visual Effects > Maya > Autodesk Maya 2019 for Novices*.

1. Choose **File > Save Scene** from the menubar to save the scene.

2. For rendering the scene, refer to Tutorial 1 of this chapter.

Tutorial 2

In this tutorial, you will create the model of a wall clock and then animate its second hand using the **Graph Editor**.   **(Expected time: 30 min)**

The following steps are required to complete this tutorial:
a. Create the project folder.
b. Create the model of a wall clock.
c. Set the animation keys and refine them.
d. Save and render the scene.

Animation

Creating the Project Folder

Create a new project folder with the name *c08_tut2* at *\Documents\maya2019* and then save the file with the name *c08tut2*, as discussed in Tutorial 1 of Chapter 2.

Creating the Model of a Wall Clock

In this section, you will create the basic model of a wall clock using the NURBS and polygon modeling methods.

1. Maximize the top-Y viewport. Choose **Create > Objects > NURBS Primitives > Circle** from the menubar and create a circle in the viewport. In the **Channel Box / Layer Editor**, enter **5** in the **Radius** edit box in the **makeNurbCircle1** node of the **INPUTS** area and press ENTER.

2. In the front-Z viewport, create another circle. In the **Channel Box / Layer Editor**, enter **0.5** in the **Radius** edit box in the **makeNurbCircle2** node of the **INPUTS** area and press ENTER.

3. Select the **Modeling** menuset from the **Menuset** drop-down list. Make sure the smaller circle is selected and then select the bigger circle by using the SHIFT key. Next, choose **Surfaces > Extrude > Options Box** from the menubar; the **Extrude Options** window is displayed. In this window, select the **At Path** radio button corresponding to the **Result Position** attribute. Also, select the **Component** radio button corresponding to the **Pivot** attribute. Now, choose the **Extrude** button; the circle is extruded.

4. Maximize the persp viewport. Choose **Windows > Editors > Outliner** from the menubar; the **Outliner** window is displayed. Select **nurbsCircle1** from the **Outliner** window. Close the **Outliner** window. Next, choose **Surfaces > Create > Planar** from the menubar; a circular NURBS surface is created. Figure 8-26 shows the base of wall clock in the persp viewport. Make sure the surface is selected. Next, choose **Curves > Reverse Direction** from the menubar; the curve direction of the selected surface is reversed.

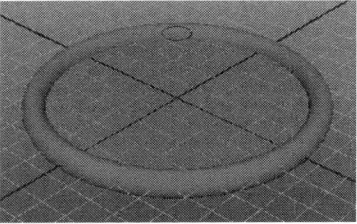

Figure 8-26 The base of wall clock

Next, you will create the text for the wall clock.

5. Choose **Create > Objects > Type** from the menubar; the settings for the text are displayed in the **Type Attributes** area of the **Attribute Editor**. In the **Type Attributes** area, select the 3D Type text and enter **3**, the **3D Type** text is replaced with 3 in the viewports.

6. Make sure **3** is selected in the viewport. In the **Channel Box / Layer Editor**, set the parameters as follows:

   Scale X : **0.14**   Scale Y : **0.14**   Scale Z : **0.14**
   Translate X: **3**   Translate Z: **0.7**   Rotate X: **-90**

7. Similarly, create numbers 9, 6, and 12. Next, arrange the text at the top-Y and persp viewports at appropriate places on the dial of the wall clock, as shown in Figure 8-27.

8. Maximize the top-Y viewport. Choose **Create > Objects > Polygon Primitives > Cylinder** from the menubar to create a cylinder at the center of the grid in the top-Y viewport. Next, set the following parameters in the **polyCylinder1** node in the **INPUTS** area of the **Channel Box / Layer Editor**:

   Radius: **0.3**   Height: **0.2**   Subdivisions Caps: **0**

9. In the top-Y viewport, choose **Create > Objects > Polygon Primitives > Cube** from the menubar to create a cube in the top-Y viewport. Next, set the following parameters in the **polyCube1** area of the **INPUTS** node of the **Channel Box / Layer Editor**:

   Width: **0.3**   Height: **0.1**   Depth: **3.5**

   Next, rename **pCube1** to *second hand* and align it with the cylinder on the wall clock, as shown in Figure 8-28.

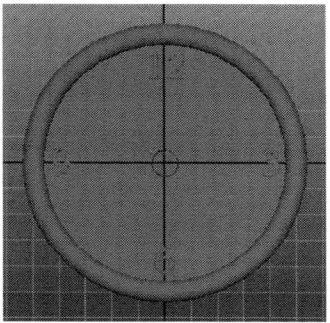
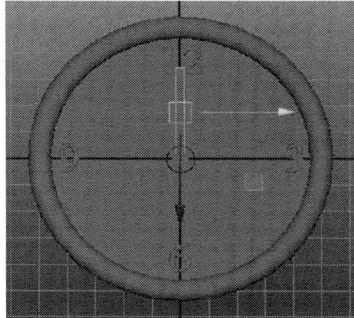

Figure 8-27 The text arranged on the clock model

Figure 8-28 The second hand of the clock

10. Make sure that the **Move Tool** is active and then press the INSERT key to display the pivot point manipulators. Next, move the pivot point of the second hand to the center of the wall clock dial. Press the INSERT key again to deactivate the manipulators.

Setting and Refining Animation Keys Using the Graph Editor

In this section, you will animate the second hand of the clock using the **Graph Editor**.

1. Select the second hand from the viewport. Next, choose **Modify > Freeze Transformations** from the menubar; the transformation values of the second hand are set to 0.

2. Set the timeslider from **1** to **5400**. Make sure that the second hand is selected at frame 1 and press S to set the key on frame 1. Now, move the timeslider to frame 30. Next, in the **Channel**

Box / Layer Editor, enter **-6** in the **Rotate Y** edit box and press S to set the animation key at frame 30.

3. Choose **Windows > Editors > Animation Editors > Graph Editor** from the menubar; the **Graph Editor** is displayed. Select **Rotate Y** from the left panel in the **Graph Editor**; the **Rotate Y** animation curve is displayed.

4. Choose **View > Infinity** from the **Graph Editor** menubar; the graph in the **Graph Editor** continues till the end. Choose **Curves > Post Infinity > Cycle with Offset** from the **Graph Editor** menubar. Next, play the animation. You will notice that the movement of the second hand is smooth. To make a strobe-like effect, you will change the tangency of the keyframe.

5. Select the **Rotate Y** animation curve from the **Graph Editor**. Next, choose **Tangents > Stepped** from the **Graph Editor** menubar to set the tangency to **Stepped** in the **Graph Editor**. Close the **Graph Editor**.

6. Preview the animation; the movement of the *second hand* becomes smooth as mechanical motion.

7. Choose the **Animation Preferences** button from the right of the **Auto keyframe toggle** button; the **Preferences** window is displayed. In this window, choose the **Settings** option from the **Categories** area. Next, select **30 fps** from the **Time** drop-down list. Choose the **Time Slider** option from the **Categories** area and make sure the **30 fps x1** option is selected in the **Playback speed** drop-down list. Next, choose the **Save** button to save the preferences.

8. Preview the animation to view the animation of the second hand in the clock.

Note
Using the steps given in this tutorial, you can create a complete clock with minute and hour hands also.

Saving and Rendering the Scene

In this section, you will save the scene that you have created and then render it. You can view the final rendered image sequence of the scene by downloading the *c08_maya_2019_rndr.zip* file from *www.cadsofttech.com*. The path of the file is as follows: *Textbooks > Animation and Visual Effects > Maya > Autodesk Maya 2019 for Novices.*

1. Choose **File > Save Scene** from the menubar to save the scene.

2. For rendering the scene, refer to Tutorial 1 of this chapter.

EXERCISE

The rendered image sequence of the scene used in the following exercise can be accessed by downloading the *c08_maya_2019_exr.zip* from *www.cadsofttech.com*. The path of the file is as follows: *Textbooks > Animation and Visual Effects > Maya > Autodesk Maya 2019 for Novices*.

Exercise 1

Download the file *c08_maya_2019_exr.zip* from *www.cadsofttech.com*. Extract the contents from the zipped file and open the scene shown in Figure 8-29. Then use the **Graph Editor**, animate the intensity of the bulb. **(Expected time: 15 min)**

Figure 8-29 The animated intensity of the bulb

Chapter 9

Rigging, Constraints, and Deformers

Learning Objectives

After completing this chapter, you will be able to:
- *Understand different types of joints*
- *Understand the parent and child relationship*
- *Use different deformers for animating an object*
- *Use different types of constraints*
- *Use the set driven keys to link objects*

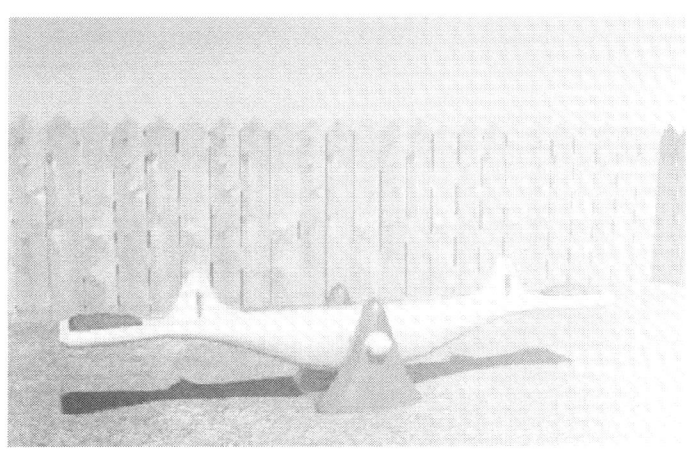

INTRODUCTION

Rigging is the process of preparing an object or a character for animation. To rig an object, you need to add bones and joints to it. Bones and joints are grouped together to form a complete skeleton. Skeleton provides support to an object in the same way as the human skeleton does to the human body. In this process, the skeleton is joined to the corresponding object by the skinning method. This method is discussed in detail later in this chapter. In this chapter, you will learn about bones and joints.

BONES AND JOINTS

Bones and joints act as the building blocks for creating a skeleton. They are visible in the viewport but cannot be rendered. Each joint may have one or more bones attached to it, as shown in Figure 9-1. Make sure the **Rigging** menuset is selected from the **Menuset** drop-down list. To create a bone, choose **Skeleton > Joints > Create Joints** from the menubar.

By default, the size of bones and joints is set to 1. To change the size of bones and joints, choose **Display > Object > Animation > Joint Size** from the menubar; the **Joint Display Scale** window will be displayed, as shown in Figure 9-2.

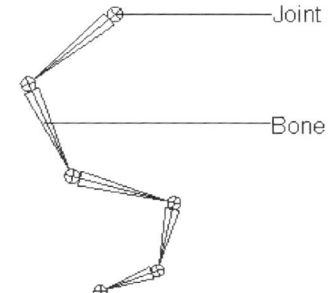

Figure 9-1 The bones and joints

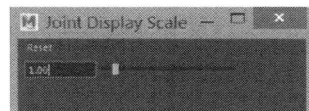

Figure 9-2 The **Joint Display Scale** window

In this window, enter the required value for the joint size in the edit box and press ENTER. Alternatively, move the slider on the right of the edit box to adjust the size of joints and bones. The **Reset** button is used to reset the value of joint size. You can also set the joint size by using the **Preferences** window. To do so, choose **Windows > Editors > Setting/Preferences > Preferences** from the menubar; the **Preferences** window will be displayed. In this window, select **Kinematics** from the **Categories** list; the **Kinematics: Kinematic Display Preferences** area will be displayed on the right in the **Preferences** window, refer to Figure 9-3. Now, enter a value in the **Joint size** edit box or move the slider on the right of the edit box to adjust the joint size in the **Inverse Kinematics** area of the **Preferences** window.

CREATING A BONE STRUCTURE

To create a bone structure in a scene, select the **Rigging** menuset from the **Menuset** drop-down list in the Status Line and activate the front-Z, side-X, or top-Y viewport. Next, choose **Skeleton > Joints > Create Joints** from the menubar and then click in the viewport; the bone will be created in the viewport. Press ENTER to exit **Joint Tool**.

Rigging, Constraints, and Deformers

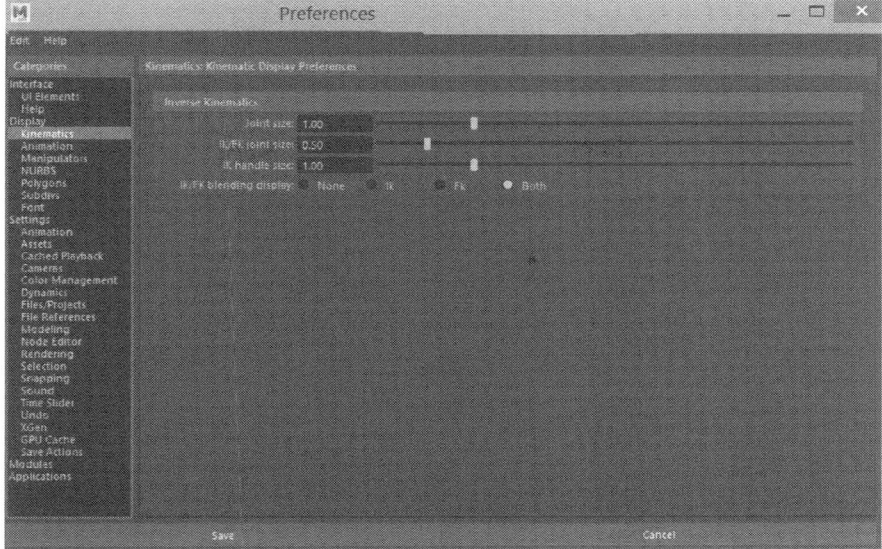

Figure 9-3 The **Preferences** *window*

To animate a joint system, you need to first set the local axes of all joints. To display the local axis of a joint, select the joint from the joint system created in the viewport and choose **Display > Objects > Transform Display > Local Rotation Axes** from the menubar; the local axes will be displayed on a single joint, as shown in Figure 9-4. Similarly, to display the local axes of all joints in a skeleton, select the topmost joint in the skeleton hierarchy and choose **Select > Hierarchy** from the menubar. Next, choose **Display > Object > Transform Display > Local Rotation Axes** from the menubar; the local axes will be displayed on all joints, as shown in Figure 9-5.

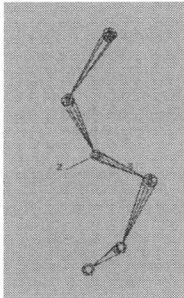 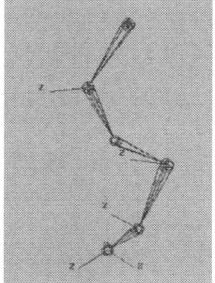

Figure 9-4 The local axes displayed on a single joint

Figure 9-5 The local axes displayed on the entire hierarchy

Types of Joints
In Maya, there are three types of joints that determine the movement of the bones attached to them. These joints are discussed next.

Ball Joint
The ball joint provides free movement to a joint in the skeleton. This type of joint can rotate about all three of its local axes freely. The human shoulder is an example of the ball joint.

Universal Joint

The universal joint provides motion to bones only in two directions. This means the joint can move freely along two axes only. The human wrist is an example of the universal joint.

PARENT-CHILD RELATIONSHIP

The parent-child relationship is the most important relationship. The parent object passes its transformations down the hierarchy chain to its children, and each child object inherits all properties of its parent. Note that a parent object can have more than one child object but not vice versa.

To understand the parent-child relationship, create two NURBS spheres in the viewport such that one sphere is larger than the other, as shown in Figure 9-6. Select the smaller sphere, press and hold the SHIFT key, and then select the larger sphere. Now, choose **Edit > Hierarchy > Parent** from the menubar; the larger sphere will become the parent of the smaller sphere. Note that the object that you select later will act as parent of the object that you selected earlier. Invoke **Move Tool** from the Tool Box and move the parent object; the child object will move along with the parent object.

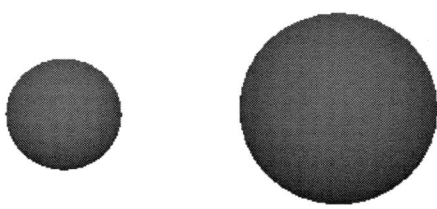

Figure 9-6 The spheres created

KINEMATICS

Kinematics is the science of motion. In the case of skeletons used in Maya, kinematics specifies the motion of bones. Kinematics is of two types: Forward and Inverse.

In Forward Kinematics (FK), the child objects are animated based on the transformations of the parent object. It is a one-way process, in which if a parent object moves, the child objects will also move. However, if a child object moves, the parent object will not move. In other words, you can use the topmost object in the hierarchy to animate the entire chain. Note that when you create a hierarchy, the Forward Kinematics is set by default.

The Inverse Kinematics (IK) is just the opposite of the Forward Kinematics. In Inverse Kinematics, you can use the object at the bottom of hierarchy to animate the entire chain. In this kinematics, if you move a child object, the objects that are higher in the hierarchy will also move accordingly.

DEFORMERS

The deformers are the tools that are used to modify the geometry of an object. You can deform any object in Maya. Various deformers in Maya are discussed next.

Blend Shape Deformer

Main menubar: Deform > Create > Blend Shape

The **Blend Shape** deformer is used to change the shape of an object into another object. The original object that is used in this process is known as the base object, and the object into which the base object gets blended is known as the target object.

To deform the shape of the polygonal base object, create a copy of the base object and modify its shape to create a target object, as shown in Figure 9-7. Now, select the target object, press and hold the SHIFT key, and then select the base object. Next, select the **Rigging** menuset from the **Menuset** drop-down list in the Status Line and choose **Deform > Create > Blend Shape** from the menubar; the blending will be done on the base object.

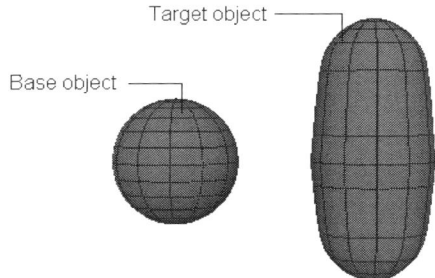

Figure 9-7 The base and target objects

Note
*You can apply the **Blend Shape** deformer on mesh objects only if they have equal number of vertices. The **Blend Shape** deformer is mainly used for creating facial expressions.*

Curve Warp Deformer

Menubar: Deform > Create > Curve Warp

The **Curve Warp Deformer** is used to stretch or animate an object along a curve. To do so, create a polygon cube (base object) and a curve. Make sure that the cube has enough subdivisions on it so that it can wrap smoothly along the curve. Select the base object and the curve using the SHIFT key. Next, select the **Modeling** menuset from the **Menuset** drop-down list in the Status Line and choose **Deform > Create > Curve Warp** from the menu bar; the polygon cube will be wrapped along the curve, refer to Figure 9-8.

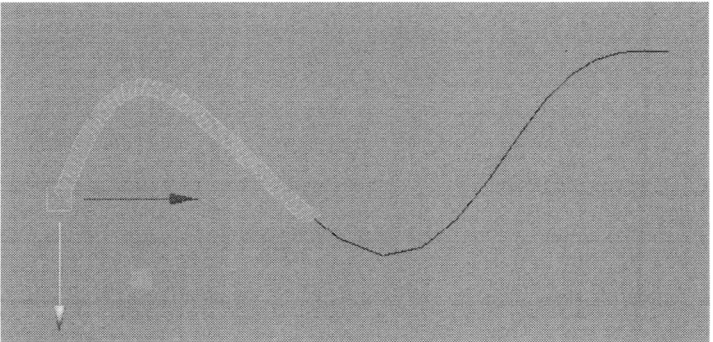

Figure 9-8 The object after applying the **Curve Warp** deformer

Lattice Deformer

Menubar: Deform > Create > Lattice

The **Lattice** deformer is used to modify an object using lattices. To modify an object using lattices, create the object in the viewport. Next, select the object and choose **Deform > Create > Lattice** from the menubar; lattice will be created around the selected object, as shown in Figure 9-9. To control the influence of lattice on the mesh, select the lattice in the viewport and enter required value in the **ffd1** area of the **OUTPUTS** node in the **Channel Box / Layer Editor**, as shown in Figure 9-10. To set the number of lattice segments, set the required values in the **S Divisions**, **T Divisions**, and **U Divisions** edit boxes of the **SHAPES** node in the **Channel Box / Layer Editor**.

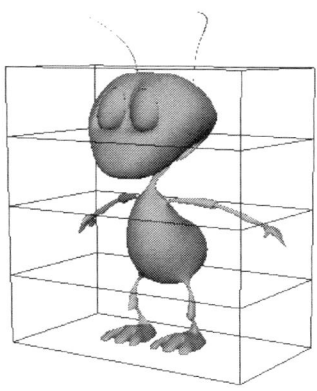

Figure 9-9 Lattice created around the selected object

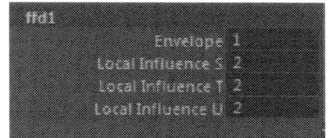

Figure 9-10 The *ffd1* area in the **OUTPUTS** node

After setting the required parameters of the lattice, you can deform the object.

Wrap Deformer

Menubar: Deform > Create > Wrap

The **Wrap** deformer is used to deform an object using NURBS surfaces, NURBS curves, or polygonal surfaces (meshes). To apply the **Wrap** deformer to an object, create a polygonal plane and add segments to it. Next, create polygonal sphere in the viewport. The polygonal sphere should be placed such that it intersects with the polygonal plane at some point. Next, invoke **Move Tool** from the Tool Box and select the polygonal sphere. Next, press and hold the SHIFT key and select the polygonal plane. Now, choose **Deform > Create > Wrap** from the menubar to apply the **Wrap** deformer. To view the deformation on sphere, select the vertices of plane and move them. You will notice changes on sphere.

ShrinkWrap Deformer

Menubar: Deform > Create > ShrinkWrap

The **ShrinkWrap** deformer is used to shrink the shape of a wrapper object according to the target object. To apply the **ShrinkWrap** deformer to an object, select it and then select a target object using the SHIFT key. Now, choose **Deform > Create > ShrinkWrap** from the menubar to apply the **ShrinkWrap** deformer.

Nonlinear Deformers

In Maya, there are different types of nonlinear deformers. Some are discussed next.

Bend Deformer

Menubar: Deform > Create > Nonlinear > Bend

The **Bend** deformer is used to bend an object along a circular arc. Figures 9-11 and 9-12 show a cylinder before and after applying the **Bend** deformer, respectively. To bend an object, select the object in the viewport. Next, choose **Deform > Create > Nonlinear > Bend** from the menubar; the **Bend** deformer will be applied to the selected object. Again, select the object in the viewport and choose **Windows > Editors > General Editors > Attribute Editor** from the menubar; the **Attribute Editor** will be displayed, as shown in Figure 9-13. Choose the **bend1** tab from the **Attribute Editor** and adjust the attributes in the **Nonlinear Deformer Attributes** area to bend the object.

Note

You should avoid changing the number of CVs, vertices, or other lattice points after applying a deformer on an object. Any change in the object will lead to a change in the functioning of that deformer.

Figure 9-11 The object before applying the **Bend** deformer

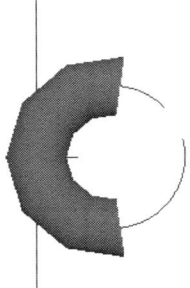

Figure 9-12 The object after applying the **Bend** deformer

Flare Deformer

Menubar: Deform > Create > Nonlinear > Flare

The **Flare** deformer is used to taper an object along two axes. Figures 9-14 and 9-15 show a cylinder before and after applying the **Flare** deformer, respectively. To taper an object using this deformer, create a NURBS cylinder in the viewport and make sure it is selected. Next, choose **Deform > Create > Nonlinear > Flare** from the menubar; the **Flare** deformer will be applied to the object. Again, select the cylinder in the viewport and choose **Windows > Editors > General Editors > Attribute Editor** from the menubar; the **Attribute Editor** will be displayed, as shown in Figure 9-16. Choose the **flare1** tab from the **Attribute Editor** and set the values for various attributes in the **Nonlinear Deformer Attributes** area to deform the object, refer to Figure 9-16.

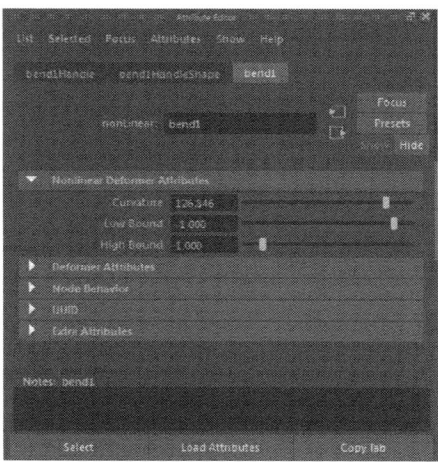

Figure 9-13 The **Nonlinear Deformer Attributes** area in the **bend1** tab

Figure 9-14 The cylinder before applying the **Flare** deformer

Rigging, Constraints, and Deformers

Figure 9-15 The cylinder modified using the **Flare** deformer

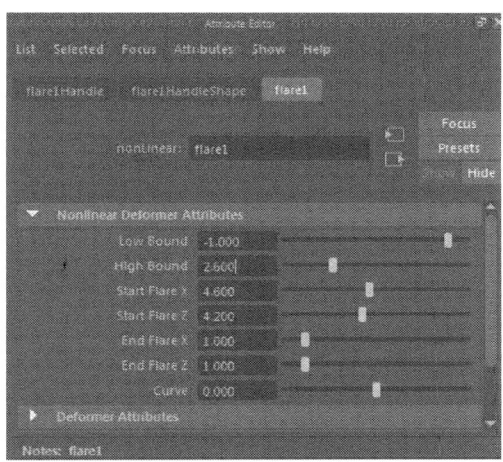

Figure 9-16 Partial view of the **Flare** deformer attributes in the **Attribute Editor**

Sine Deformer

Menubar: Deform > Create > Nonlinear > Sine

The **Sine** deformer is used to deform an object in the shape of a sine wave. Figures 9-17 and 9-18 show a cylinder before and after applying the **Sine** deformer, respectively. To apply this deformer, select an object in the viewport and then choose **Deform > Create > Nonlinear > Sine** from the menubar; the **Sine** deformer will be applied to the object and the **Attribute Editor** will be displayed, refer to Figure 9-19. Next, choose the **sine1** tab from the **Attribute Editor** and set the values of various attributes in the **Nonlinear Deformer Attributes** area to deform the object.

Figure 9-17 The cylinder before applying the **Sine** deformer

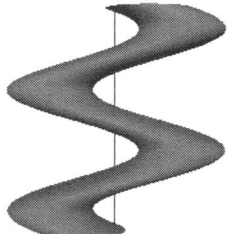

Figure 9-18 The cylinder after applying the **Sine** deformer

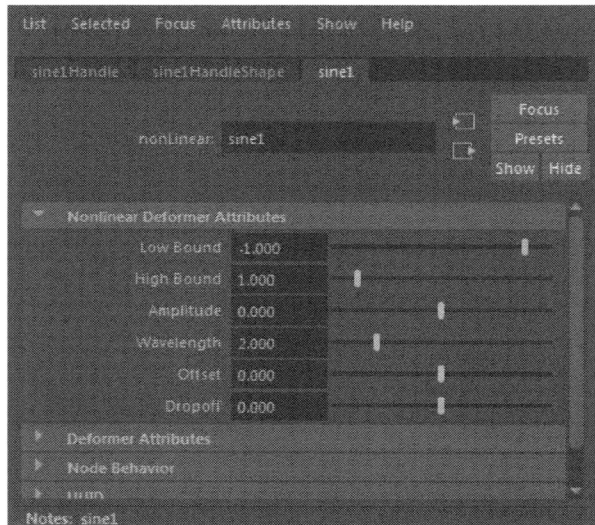

*Figure 9-19 Partial view of the **Sine** deformer attributes in the **Attribute Editor***

Twist Deformer

| **Menubar:** | Deform > Create > Nonlinear > Twist |

The **Twist** deformer is used to twist an object about an axis. To apply this deformer, select an object in the viewport and choose **Deform > Create > Nonlinear > Twist** from the menubar; the **Twist** deformer will be applied to the object. Now, select the object again from the viewport and choose **Windows > Editors > General Editors > Attribute Editor** from the menubar; the **Attribute Editor** will be displayed. Choose the **twist1** tab from the **Attribute Editor** and set the values of various attributes in the **Nonlinear Deformer Attributes** area to deform the object, as shown in Figure 9-20.

Rigging, Constraints, and Deformers

*Figure 9-20 The **Nonlinear Deformer Attributes** area in the **twist1** tab*

Wave Deformer

| **Menubar:** | Deform > Create > Nonlinear > Wave |

The **Wave** deformer is used to propagate waves on an object in the X and Z directions. Figures 9-21 and 9-22 show a plane before and after applying the **Wave** deformer, respectively.

*Figure 9-21 The plane before applying the **Wave** deformer*

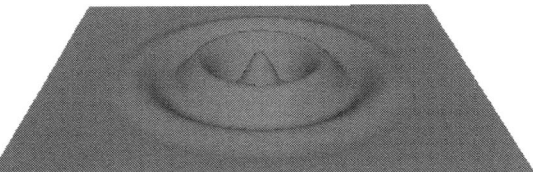

*Figure 9-22 The plane after applying the **Wave** deformer*

To apply the **Wave** deformer, select an object in the viewport and then increase the number of segments on it from the **Channel Box / Layer Editor**. Next, choose **Deform > Create > Nonlinear > Wave** from the menubar; the **Wave** deformer will be applied to the selected object. Select the object again and choose **Windows > Editors > General Editors > Attribute Editor** from the menubar; the **Attribute Editor** will be displayed. Next, choose the **wave1** tab from the **Attribute Editor** to deform the selected object as desired. The attributes of the **Wave** deformer are similar to those of the **Sine** deformer and are shown in Figure 9-23.

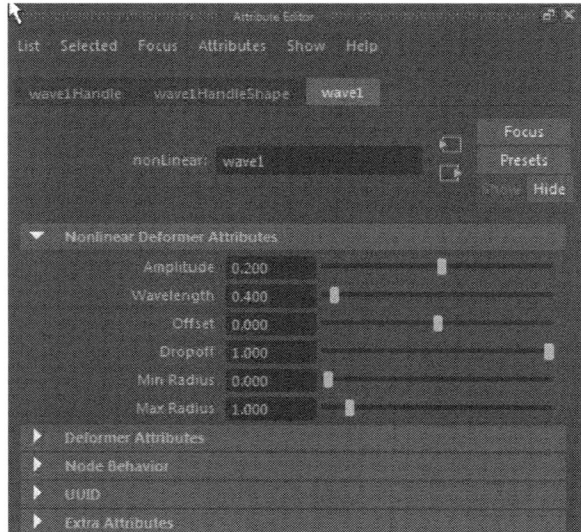

*Figure 9-23 Partial view of the **Wave** deformer attributes in the **Attribute Editor***

APPLYING CONSTRAINTS

Constraints are used to restrict the motion of an object to a particular mode by specifying their limits. Some constraints in Maya are discussed next.

Parent Constraint

| Menubar: | Constrain > Create > Parent |

The Parent constraint is used to relate the orientation of one object with the other object such that both of them follow the parent-child relationship. To apply this constraint, create two objects in the viewport. Select one object, press and hold the SHIFT key, and then select the other object. Next, choose **Constrain > Create > Parent** from the menubar to apply the Parent constraint to the selected objects. Change the position of the parent object; the objects follow the parent-child relationship. The Parent constraint is different from the Point and Orient constraints. When an object is rotated using the Point or Orient constraint, the constrained object rotates about its local axis. Whereas in case of the Parent constraint, the constrained object rotates with respect to the world axis.

Point Constraint

| Menubar: | Constrain > Create > Point |

The Point constraint is used to restrict the movement of an object such that the constrained object follows the movement of another object. To apply this constraint, create two cubes of different sizes in the viewport. Next, select the **Rigging** menuset from the Status Line. Now, select one cube, and press and hold the SHIFT key to select another cube. Next, choose **Constrain > Create > Point** from the menubar to coordinate the motion of one cube with another cube. The object selected first controls the movement of the object selected later. On applying the **Point** constraint, the objects may overlap when they are moved. To avoid this situation, choose

Constrain > Create > Point > Option Box from the menubar; the **Point Constraint Options** window will be displayed, as shown in Figure 9-24. The **Offset** attribute in this window is used to set the distance between the two selected objects. Enter the required values in the **Offset** edit boxes and choose the **Add** button from the window; the Point constrain will be applied to the selected object.

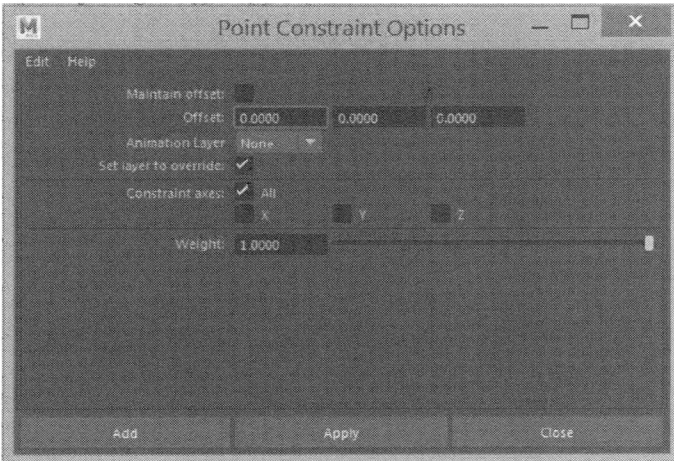

*Figure 9-24 The **Point Constraint Options** window*

Note
The working of a constraint is opposite to that of the parent-child relationship. In a parent-child relationship, the object selected later acts as the parent, but in case of constraints, the object selected first acts as the parent of the object selected later.

SET DRIVEN KEY

The **Set Driven Key** is used to link the attribute of one object to another object. When you set the driven key, you need to specify a driver value and a driven attribute value. In such a case, the value of the driven attribute is locked to the corresponding value of the driver attribute. Therefore, a change in the driver attribute will change the value of the driven attribute as well. Select the **Animation** menuset from the **Menuset** drop-down list. To set a driven key, choose **Key > Set > Set Driven Key > Set** from the menubar; the **Set Driven Key** window will be displayed, as shown in Figure 9-25. Select the object from the viewport that you want to set as the driver and then choose the **Load Driver** button; the name of the object with its attributes will be displayed in the **Driver** area of the **Set Driven Key** window.

*Figure 9-25 The **Set Driven Key** window*

TUTORIAL

All the files used in this tutorial can be downloaded from the CADSoft website (*www.cadsofttech.com*). These files are compressed in zip file format and are required to be extracted before using them in the tutorials. The path of the files is as follows: *Textbooks > Animation and Visual Effects >Maya > Autodesk Maya 2019 for Novices*

Tutorial 1

In this tutorial, you will create the bone structure of a human leg, as shown in Figure 9-26, using **Joint Tool**. **(Expected time: 15 min)**

The following steps are required to complete this tutorial:

a. Create a project folder.
b. Create the bone structure of the leg.
c. Apply IKs to the bone structure.
d. Create the reverse foot setup.
e. Create the pole vector.
f. Save the scene.

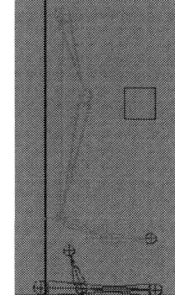

Figure 9-26 The bone structure of a human leg

Creating a Project Folder

Create a new project folder with the name *c09_tut1* at *\Documents\maya2019* and then save the file with the name *c09tut1*, as discussed in Tutorial 1 of Chapter 2.

Creating the Bone Structure of Leg

In this section, you will create the bone structure of a human leg.

1. Maximize the side-X viewport. Select the **Rigging** menuset from the **Menuset** drop-down list in the Status Line. Next, choose **Skeleton > Joints > Create Joints** from the menubar. Next, create the bone structure in the viewport and press the ENTER key, refer to Figure 9-27. (It is recommended that you start the structure from the *Hip* joint).

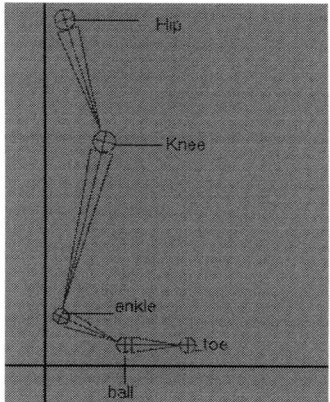

Figure 9-27 The bone structure

2. Select the *Hip* joint from the bone structure, refer to Figure 9-27; the entire bone structure

Rigging, Constraints, and Deformers

is selected. In the **Channel Box / Layer Editor**, click on the default joint name and rename it as *left_Hipjoint*. Similarly, name other joints as given below:

Joints	Names
Knee	*left_Kneejoint*
Ankle	*left_Anklejoint*
Ball	*left_Balljoint*
Toe	*left_Toejoint*

Note

Naming the joints of a character is very important because it helps you while animating and skinning the character. Use the word 'left' to name the left body joints and 'right' to name the right body joints.

Applying IKs to the Bone Structure

In this section, you will apply IKs to the bone structure.

1. Choose **Skeleton > Ik > Create IK Handle** tool from the menubar.

2. Select the *left_Hipjoint* joint and then the *left_Anklejoint* joint in the viewport; an IK handle is created between these two joints, as shown in Figure 9-28. Rename the IK handle as *ikhandle_ankle* in the **Channel Box / Layer Editor**.

3. Similarly, create other IK handles between the *left_Anklejoint* and *left_Balljoint*, and also between the *left_Balljoint* and *left_Toejoint*, as shown in Figure 9-29. Rename the IK handles as *ikhandle_ball* and *ikhandle_toe*.

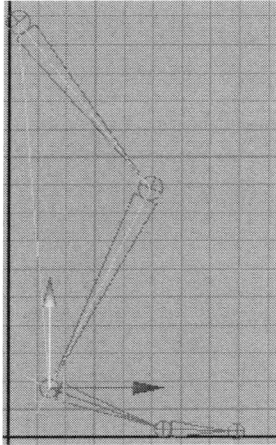 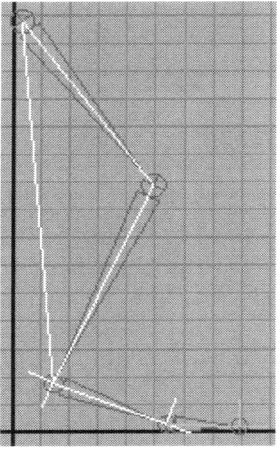

Figure 9-28 An IK Handle created between joints

Figure 9-29 The IK Handles

Tip
*To adjust the joints of a bone structure, press and hold the d key and invoke **Move Tool**. Once the tool is invoked, you can move the joints to adjust the bone structure. You can also resize the bone structure. To do so, choose **Display > Object > Animation > Joint Size** from the main menubar; the **Joint Display Scale** window will be displayed. Now, you can adjust the joint size as required.*

Creating the Reverse Foot Setup

In this section, you need to create the reverse foot setup that will provide control to the movement of the leg.

1. In the side-X viewport, choose **Skeleton > Joints > Create Joints** from the menubar; the **Create Joints** tool is activated. Next, use this tool to create reverse foot setup, as shown in Figure 9-30. Rename 1, 2, 3, and 4 joints as *rf_leftheeljoint*, *rf_lefttoejoint*, *rf_leftballjoint*, and *rf_leftanklejoint*, respectively, refer to Figure 9-30.

2. Invoke **Move Tool**. Next, select *ikhandle_ankle*, press SHIFT and then select the joint 4, refer to Figure 9-31. Now, press P in the keyboard to make the joint *rf_leftanklejoint* of the reverse foot as parent of *ikhandle_ankle*.

3. Similarly, make the joint *rf_lefttoejoint* as the parent of the *ikhandle_toe* and the joint *rf_leftballjoint* as the parent of the *ikhandle_ball*. Now, hold the reverse foot setup and move the foot as required.

Creating the Pole Vector

In this section, you will create the pole vector to control the movement of the knee joint.

1. Select ikhandle_ankle in the side-X viewport. In the **Attribute Editor**, choose the **ikhandle_ankle** tab and expand the **IK Solver Attributes** area. In this area, select the **Rotate-Plane Solver** option from the **IK-Solver** drop-down list, as shown in Figure 9-31. Next, enter **0**, **1**, **0** in the **Pole Vector** edit boxes.

2. Create a polygon cube in the side-X viewport. Invoke Move Tool from the Tool Box and align the cube near the knee in all viewports.

3. Select the polygon cube, press SHIFT, and then select *ikhandle_ankle*. Next, choose **Constrain > Create > Pole Vector** from the menubar; the polygon cube pole vector of the knee is created, as shown in Figure 9-32. Move the cube left and right; the knee joint will move accordingly.

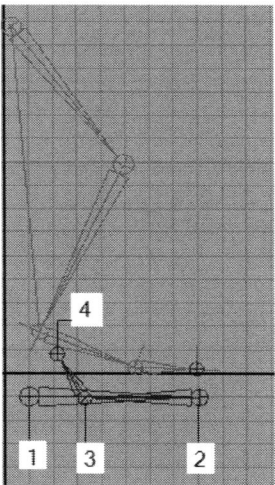

Figure 9-30 The reverse foot setup

4. Maximize the persp viewport and invoke Move Tool from the Tool Box to check the movement of the foot by using IKs and the pole vector.

Saving the Scene

In this section, you need to save the scene that you have created.

1. Choose **File > Save Scene** from the menubar to save the scene.

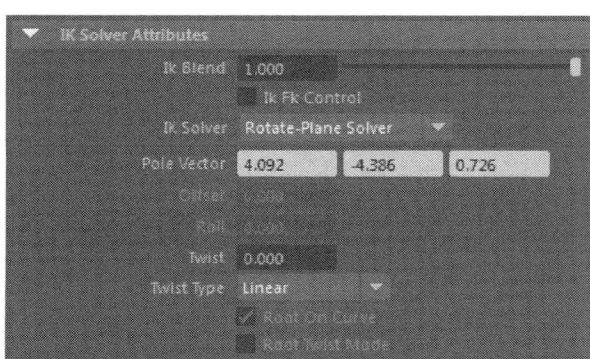
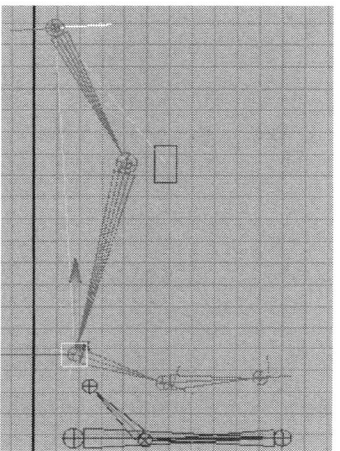

Figure 9-31 Selecting **Rotate-Plane Solver** from the **IK Solver** drop-down list

Figure 9-32 The pole vector constrain

EXERCISES

The image sequence of the scenes used in the following exercises can be accessed by downloading the *c09_maya_2019_exr.zip* from *www.cadsofttech.com*. The path of the file is as follows: *Textbooks > Animation and Visual Effects > Maya > Autodesk Maya 2019 for Novices*.

Exercise 1

Create a simple toy car and then use the **Set Driven Key** method to set keys for the doors of the toy car. **(Expected time: 30 min)**

Exercise 2

Create a pencil stand with a pencil in it. Next, apply texture to the model. Apply the **Lattice** deformer to the pencil. Next, use the keyframe animation technique to make the pencil jump out of the pencil stand, refer Figure 9-33. **(Expected time: 45 min)**

Figure 9-33 The pencil jumping out of the pencil stand

Chapter 10

Paint Effects

Learning Objectives

After completing this chapter, you will be able to:
- *Use the Content Browser window*
- *Render the paint effect strokes*
- *Use shadow effects*

INTRODUCTION

In Autodesk Maya, you can create realistic natural objects such as trees, plants, rain, and so on by using paint effects. The paint effects help you to paint a scene by using a mouse or a tablet. Different brushes are used to create effects such as rain, thunder, storm, and so on. You can also animate paint effects to create natural motion. All these paint effects and brushes are available in the **Content Browser** window. The **Content Browser** window is discussed next.

WORKING WITH THE Content Browser WINDOW

| **Menubar:** | Windows > Editors > General Editors > Content Browser |

The **Content Browser** window comprises of preloaded animation clips, default brushes, shader libraries or texture libraries, and so on. To open this window, choose **Windows > Editors > General Editors > Content Browser** from the menubar. The **Content Browser** window is displayed in Figure 10-1. There are various nodes in this window such as **Animation**, **FX**, **Modeling**, **Paint Effects**, and so on. When you choose a particular node, its corresponding nodes will be displayed in the right pane of the **Content Browser** window. For example, when you choose the **Paint Effects** node in the left pane of the **Content Browser** window, various paint stroke nodes will be displayed in the right pane.

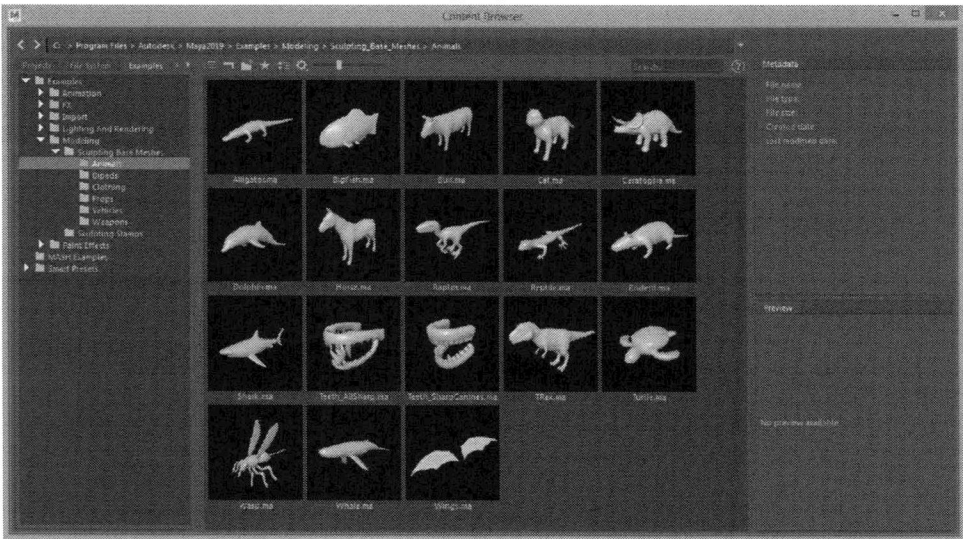

Figure 10-1 The **Content Browser** *window*

Creating Objects

You can create a realistic object such as trees, buildings, and so on using the **Content Browser** window. For example, to create a tree, choose the **Paint Effects > Trees** node at the left pane of the **Content Browser** window; various tree nodes will be displayed in the right pane of the **Content Browser** window. Now, choose the **oakLimb.mel** paint stroke from the displayed options; the shape of the cursor will change into a pencil. Next, activate the top-Y viewport. Press and hold the left mouse button and drag the cursor to create the tree mesh. Next, activate the persp viewport and render the view to get the output shown in Figure 10-2.

Paint Effects

Figure 10-2 *The rendered image*

You can also edit the paint stroke created in the viewport. To do so, select the paint stroke created in the viewport; the name of the selected paint stroke will be displayed in the **INPUTS** area of the **Channel Box / Layer Editor**. Click on the paint stroke name to expand its attributes. You can now modify the selected paint stroke as per your requirement using the attributes in the **Channel Box / Layer Editor**.

WORKING WITH THE Paint Effects WINDOW

You can also draw the paint strokes in the viewport. To do so, choose **Windows > Editors > Modeling Editors > Paint Effects** from the menubar; the **Paint Effects** window will be displayed, as shown in Figure 10-3.

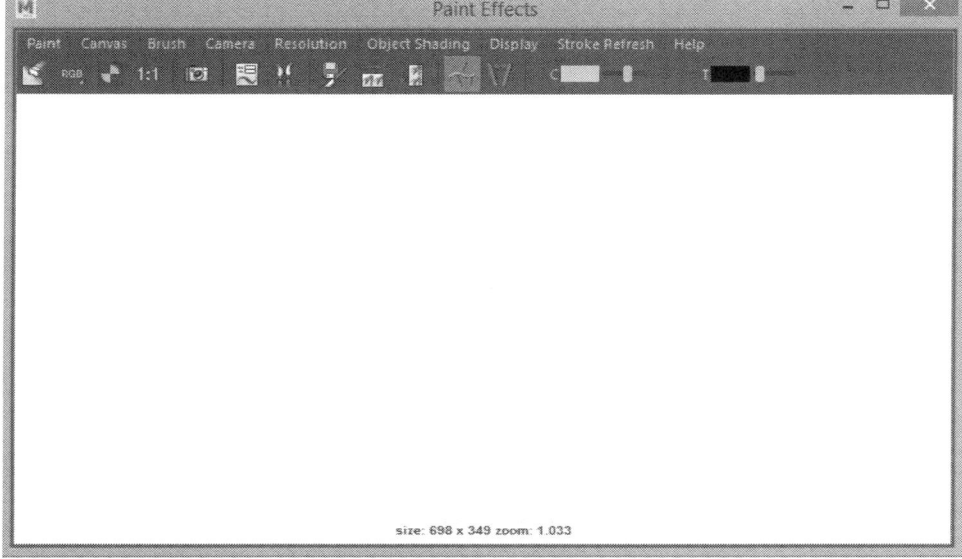

Figure 10-3 *The **Paint Effects** window*

The **Paint Effects** window has its own menubar and toolbar, as shown in Figure 10-4. It consists of menus such as **Paint**, **Canvas**, **Brush**, **Cameras**, and so on, refer to Figure 10-4. The tools in this toolbar of the **Paint Effects** window are used to create different effects by using the paint strokes. To invoke a paint stroke brush, choose the **Get brush** tool from the toolbar, refer to Figure 10-4; the **Content Browser** window will be displayed. Choose the required paint brush stroke from the **Content Browser** window and then paint the stroke in the **Paint Effects** window. You can also edit the attributes of the selected paint brush stroke by using the options in this window. To do so, select the paint stroke from the **Content Browser** window and choose the **Edit template brush** tool from the toolbar of the **Paint Effects** window; the **Paint Effects Brush Settings** window will be displayed. Alternatively, press CTRL+B to invoke the **Paint Effects Brush Settings** window, as shown in Figure 10-5. Some of the basic options of this window are discussed next.

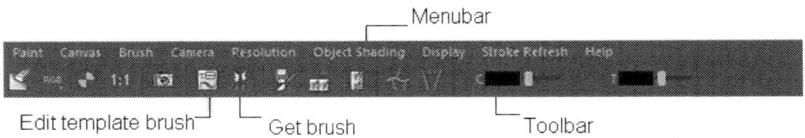

Figure 10-4 *The menubar and toolbar of the* **Paint Effects** *window*

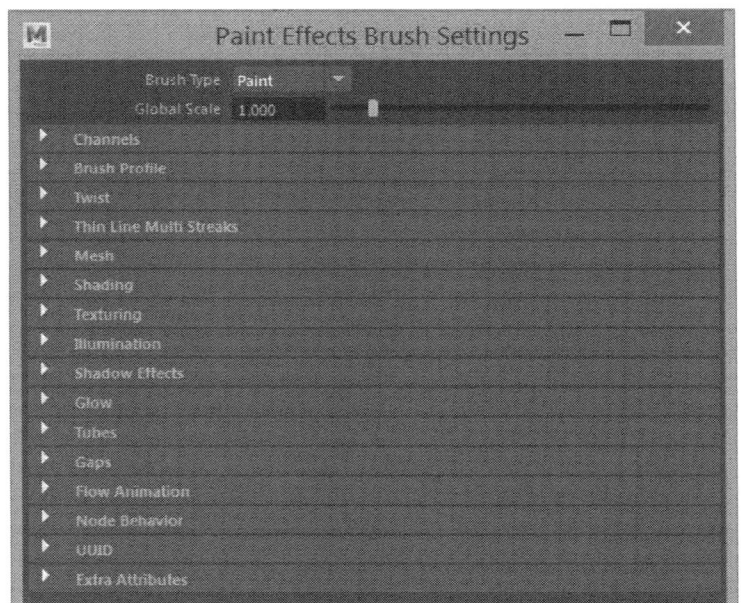

Figure 10-5 *The* **Paint Effects Brush Settings** *window*

Brush Type

The options in the **Brush Type** drop-down list are used to select the type of brush you want to use. The shape used by the brushes is defined by the brush attributes. The **Paint** brush type applies the paint to stroke path according to the brush attributes you have set. The **Smear** brush type distorts the stroke (paint) already applied to the canvas or scene. If you have enabled fake shadows from the **Shadow Effects** area of the **Paint Effects Brush Settings** window, the shadows will smear as well. The **Blur** brush type is used to soften the paint already applied to the canvas. The **Erase** brush type removes the paint from the canvas, revealing the color of the canvas. The **ThinLine** brush type allows large numbers of brush stroke quickly than the **Paint** brush type. The **Mesh** brush type is used to create accurate conical geometry with textures that correctly map on the surface.

Global Scale

The **Global Scale** attribute is used to change the value of the brush attributes by a common factor so that you can paint the same stroke in different sizes. When you specify a value for this attribute, the paint effect is scaled uniformly by this value. The default value of this option is 1. Figures 10-6 and 10-7 show an object created by specifying two different values for the **Global Scale** attribute.

Figure 10-6 Object with the **Global Scale** value = 1

Figure 10-7 Object with the **Global Scale** value = 2

Tip
*You can interactively specify a value for the **Global Scale** attribute. To do so, press B + LMB and then drag to the left or right.*

Channels

Generally, a rendered image consists of three channels: red, green, and blue. These channels represent amount of red, green, and blue colors in the image. Some images may also contain some additional channels such as alpha, mask, and depth. The depth channel is also referred to as Z depth or Z buffer channel. These additional channels are used extensively when artwork is composited in a compositing software such as Fusion or Nuke. By default, paint effects contain three color channels (RGB) and an alpha channel. The attributes in the **Channels** area are used to specify the depth, color, and alpha settings. In the **Paint Effects Brush Settings** window, click on the arrow on the left of the **Channels** area to expand it, if not already expanded. On doing so, the **Depth**, **Modify Depth**, **Modify Color**, and **Modify Alpha** check boxes will be displayed. Select the **Depth** check box to create a depth channel. You will notice that brush strokes in the scene appear more natural and realistic. Select the **Modify Depth** check box to paint the depth channel. Select the **Modify Color** and **Modify Alpha** check boxes to paint the color and alpha channels, respectively.

Brush Profile

The attributes in the **Brush Profile** area are used to set the brush settings. On expanding this area, various options will be displayed, as shown in Figure 10-8. Some of these options are discussed next.

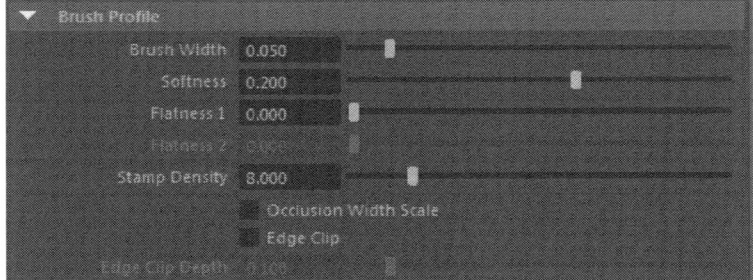

Figure 10-8 The **Brush Profile** area

Brush Width

The **Brush Width** attribute is used to define the width of the brush. The brush width is defined by the outline of the paint effect generated. Enter a value in the **Brush Width** edit box or move the slider on its right to set the value of the brush width.

Softness

The **Softness** attribute is used to define the blurriness on the edges of the stroke path. The higher the softness value, the more blurred will be the edges. Refer to Figures 10-9 and 10-10 for variations in the **Softness** value.

Figure 10-9 Paint stroke with **Softness** = 0   *Figure 10-10* Paint stroke with **Softness** = 1

Flatness 1 and Flatness 2

These attributes are used to flatten the paint strokes along the stroke path or to flat each tube at its base and tip. If you are drawing simple strokes, the **Flatness 1** option defines the flatness of the paint strokes along the stroke paths. However, if you are drawing tubes, the **Flatness 1** and **Flatness 2** defines the flatness of each tube at its base and tip. Figures 10-11 and 10-12 show the paint strokes created by using different values of flatness.

Figure 10-11 Paint stroke with **Flatness 1** = 0   *Figure 10-12* Paint stroke with **Flatness 1** = 0.5

Paint Effects

Stamp Density
When you draw strokes on the canvas, the paint is applied to strokes in overlapping stamps. If a stroke has no tube, the stamps will be applied along the stroke path. However, if a stroke has tubes, the stamps will be applied along the tube path. The **Stamp Density** attribute defines the number of stamps to be applied along the path. The **Stamp Density** attribute is related to the **Brush Width** attribute. For example, if you specify a value of 3 for the **Brush Width** attribute and a value of 6 to the **Stamp Density** option, there will be 8 stamps in every 3 units of path.

Occlusion Width Scale
Select this check box if you are using a toon shader. This option reduces the stamp size if foreground objects are overlapping the stamp.

Edge Clip and Edge Clip Depth
Select the **Edge Clip** check box to render 3D strokes as flat 2D strokes. It gives an illusion as if the strokes are directly painted on the texture of a surface. The **Edge Clip Depth** option controls the distance between the surface and a point beyond which the stroke will become visible.

Twist
The attributes in the **Twist** area are used to twist tubes around their own axis as they grow. When you expand the **Twist** area, some more attributes will be displayed, as shown in Figure 10-13. These options are discussed next.

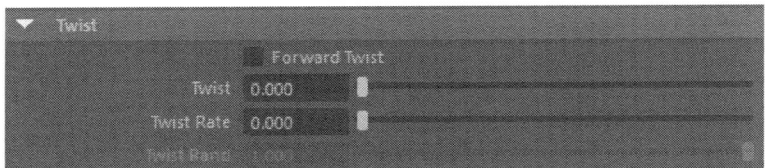

*Figure 10-13 The **Twist** area*

Forward Twist
When this check box is selected, the flat sides of tubes and textures always face the camera.

Twist
The **Twist** attribute defines the initial value of the twist. This attribute is affected by the **Flatness 1** and **Flatness 2** options. Twist is noticeable in the strokes only if the value of the **Flatness 1** and **Flatness 2** options is greater than 0.

Twist Rate
This attribute controls the strength of twist along the length of the strokes. Twist will be only noticeable in the strokes if the **Flatness 1** and **Flatness 2** options' value is greater than 0. Figure 10-14 shows the paint stroke with **Tube Rate** value set to **0** and Figure 10-15 shows the paint stroke with **Tube Rate** value set to **3**.

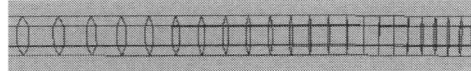   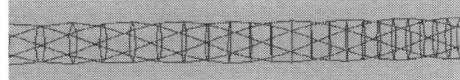

*Figure 10-14 Paint stroke with **Tube Rate** = 0*   *Figure 10-15 Paint stroke with **Tube Rate** = 3*

Twist Rand

This attribute is used to define the randomness applied to the twist.

Mesh

The attributes in the **Mesh** area are used to define the mesh brush. On expanding this area, the **Mesh** area will be displayed, as shown in Figure 10-16.

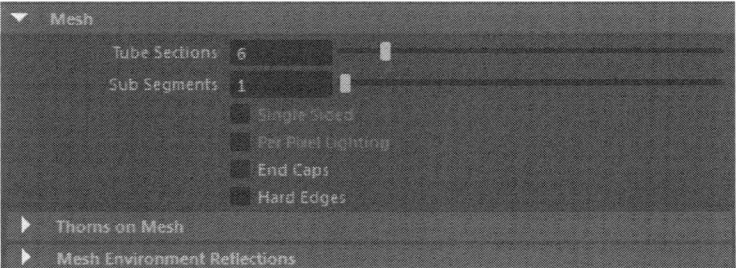

Figure 10-16 The **Mesh** area

Thorns on Mesh

The attributes in the **Thorns on Mesh** area are used to add branch thorns on a mesh object. By default, the attributes in this area are inactive. To activate them, select the **Mesh** brush type from the **Brush Type** drop-down list and then expand the **Thorns on Mesh** area. Next, choose the **Branch Thorns** check box to activate the remaining options, refer to Figure 10-17. Note that the thorns are not visible in the viewport. They are visible only at the time of rendering. Figures 10-18 and 10-19 show the paint strokes before and after using the options of the **Thorns on Mesh** area. You can modify the values of density, elevation, length, base width, tip width, specular, and so on for thorns in this area to get the desired result.

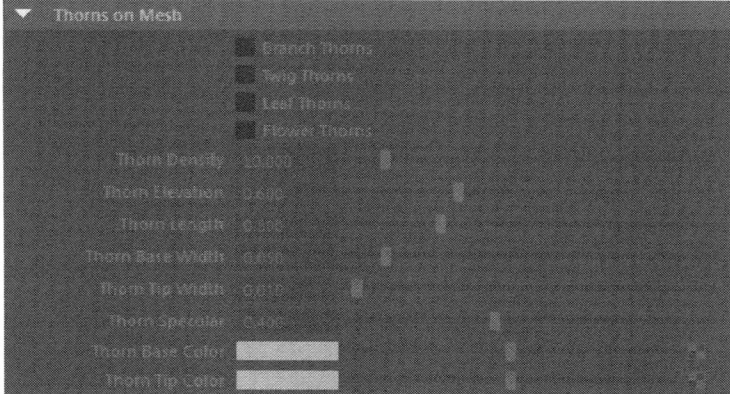

Figure 10-17 The expanded **Thorns on Mesh** area

Paint Effects

Figure 10-18 Paint stroke before using the **Thorns on Mesh** attribute

Figure 10-19 Paint stroke after using the **Thorns on Mesh** attribute

Shading

The attributes in the **Shading** area are used to define the shading of the brush strokes. These options will be displayed on expanding this area, as shown in Figure 10-20, and are discussed next.

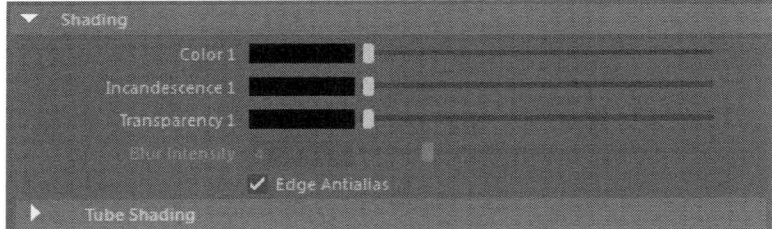

Figure 10-20 The **Shading** area

Illumination

The attributes in this area are used to change the appearance of the brush strokes by using the lighting, refer to Figure 10-21. Select the **Illuminated** check box to affect the appearance of the stroke. If you clear this check box, no shaded areas or specularity will be visible on the paint strokes even if there are lights in the scene. The **Real Lights** check box will be active only, if you have selected the **Illuminated** check box. When the **Real Lights** check box is selected, the lights in the scene determine the position of shading and specular highlights. If this check box is not selected, a directional paint effects light will be used. You can define its direction by using the **Light Direction** attribute but you cannot change any other attribute of the directional light. Figures 10-22 and 10-23 show an object before and after using the options of the **Illumination** area.

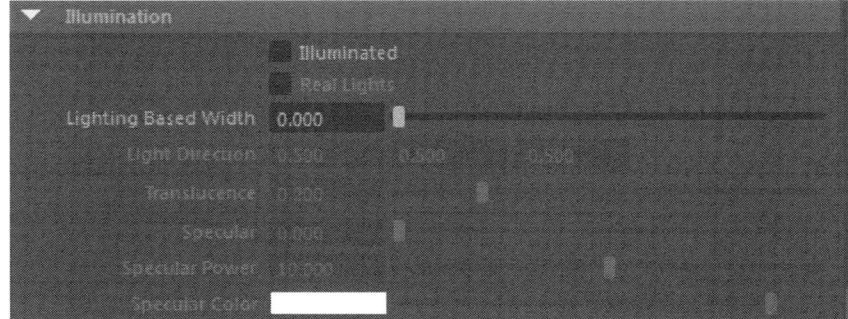

Figure 10-21 The **Illumination** area

Figure 10-22 Paint stroke with the **Illuminated** check box cleared

Figure 10-23 Paint stroke with the **Illuminated** check box selected

Shadow Effects

The attributes in the **Shadow Effects** area are used to apply shadow effect to brush strokes. To apply this effect to brush strokes, expand the **Shadow Effects** area, refer to Figure 10-24, and then adjust the attributes as required to assign the shadow effect to the brush strokes. Some of the attributes in the **Shadow Effects** area are explained next.

Paint Effects

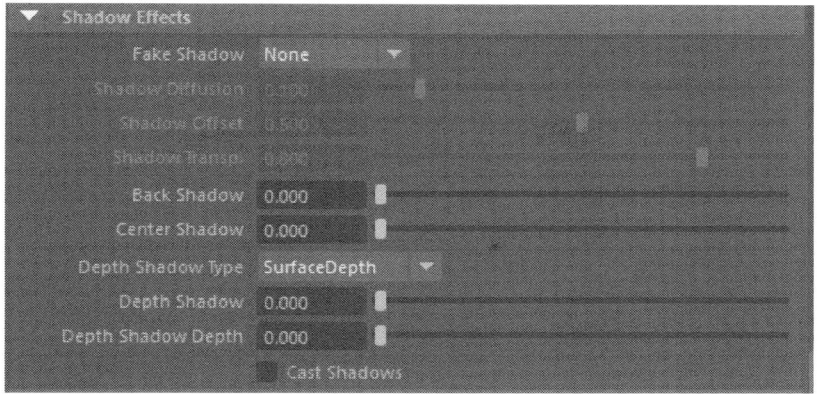

*Figure 10-24 The **Shadow Effects** area*

Fake Shadow

The options in the **Fake Shadow** drop-down list are used to create fake shadows for the brush strokes. It has three options: **None**, **2D Offset**, and **3D Cast**. The **2D Offset** option is used to create a drop shadow like effect. The **3D Cast** option is used to create a flat surface below the stroke and then to cast shadow on that imaginary surface.

Shadow Diffusion

The **Shadow Diffusion** attribute is used to control the softness of fake shadows in a scene, refer to Figures 10-25 and 10-26.

*Figure 10-25 Paint stroke with the **Shadow Diffusion** value = 0*

*Figure 10-26 Paint stroke with the **Shadow Diffusion** value = 1*

Shadow Offset

The **Shadow Offset** attribute is used to control the distance between the shadow and the casting stroke. This attribute is inactive by default. To activate this attribute, select **2D Offset** from the **Fake Shadow** drop-down list. Next, set the offset distance in the **Shadow Offset** edit box or move the slider on its right as required. Figures 10-27 and 10-28 show an object with different shadow offset values.

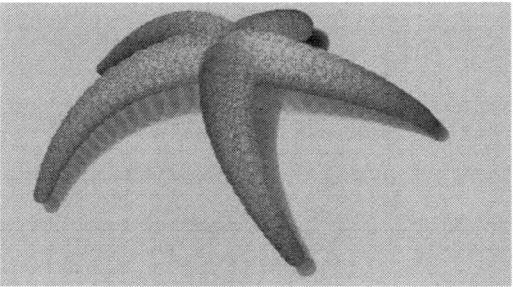 

Figure 10-27 Paint stroke with the **Shadow Offset** value = 0.5

Figure 10-28 Paint stroke with the **Shadow Offset** value = 1

Shadow Transp

The **Shadow Transp** attribute is used to specify the value of transparency of the shadow of the paint stroke. Higher the transparency value, lighter will be the shadow effect and vice versa. Figures 10-29 and 10-30 show an object with different values of the **Shadow Transp** attribute.

Figure 10-29 Paint stroke with the **Shadow Transp** value = 0

Figure 10-30 Paint stroke with the **Shadow Transp** value = 0.8

Glow

The attributes in the **Glow** area used to add standard glow to paint strokes, refer to Figure 10-31. The **Glow** attribute defines the brightness of the glow. Higher the value of the **Glow** option, more will be the glow, as shown in Figures 10-32 and 10-33. The **Glow Color** attribute defines the color of the standard glow. There will be no glow if you set the **Glow Color** attribute to black. The **Glow Spread** attribute controls the halo around the paint strokes. The **Shader Glow** attribute controls the brightness of the shader glow and is more realistic than the standard glow.

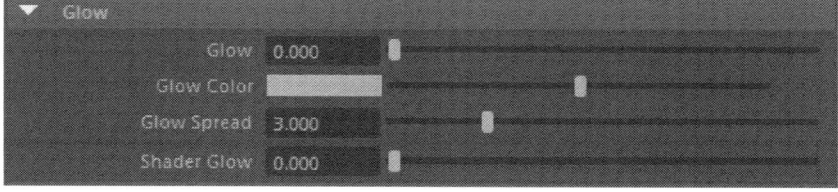

Figure 10-31 The **Glow** area

Paint Effects

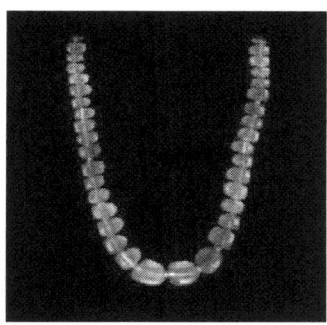

Figure 10-32 Paint stroke with the **Glow** value = 0

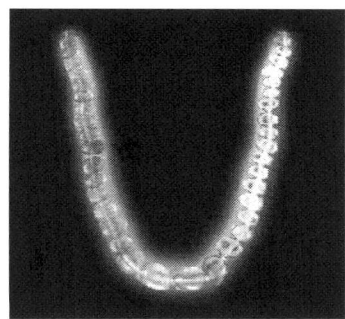

Figure 10-33 Paint stroke with the **Glow** value = 0.2

TUTORIAL

All the files used in this tutorial can be downloaded from the CADSoft website (*www.cadsofttech.com*). These files are compressed in zip file format and are required to be extracted before using them in the tutorials. The path of the files is as follows: *Textbooks > Animation and Visual Effects >Maya > Autodesk Maya 2019 for Novices*

Tutorial 1

In this tutorial, you will create a street scene, as shown in Figure 10-34, by using the paint effects. **(Expected time: 20 min)**

The following steps are required to complete this tutorial:

a. Create a project folder.
b. Download the texture file.
c. Create a road for the street scene.
d. Create buildings.
e. Create clouds.
f. Create lights.
g. Save and render the scene.

Figure 10-34 A street scene

Creating a Project Folder

Create a new project folder with the name *c10_tut1* at *\Documents\maya2019* and then save the file with the name *c10tut1*, as discussed in Tutorial 1 of Chapter 2.

Downloading the Texture File

In this section, you need to download the texture file.

1. Download the *c10_maya_2019_tut.zip* file from *www.cadsofttech.com*. The path of the file is as follows: *Textbooks > Animation and Visual Effects > Maya > Autodesk Maya 2019 for Novices*.

2. Extract the contents of the zip file to the *Documents* folder. Open Windows Explorer and then browse to *\Documents\c10_maya_2019_tut*. Next, copy *roadtexture.jpg* to *\Documents\maya2019\ c10_tut1\sourceimages*.

Creating a Road for the Street Scene

In this section, you need to create a road for the street scene by using polygon primitives.

1. Maximize the top-y viewport and choose **Create > Objects > Polygon Primitives > Plane** from the menubar.

2. In **Channel Box / Layer Editor**, set the parameters of **polyPlane1** in the **INPUTS** area, as shown in Figure 10-35.

Figure 10-35 Setting the parameters of polyPlane1

3. Choose **Windows > Editors > Rendering Editors > Hypershade** from the menubar; the **Hypershade** window is displayed. Choose **Maya > Lambert** from the **Create** area of this window; the **lambert2** shader is created in the Work Area of the **Hypershade** window.

Paint Effects

4. Press and hold CTRL and double-click on the **lambert2** shader; the **Rename node** window is displayed. Enter **road** in the **Enter new name** edit box and choose the **OK** button; the shader is renamed to *road*.

5. Select the plane in the viewport and then press and hold the right mouse button over the plane; a marking menu is displayed. Next, choose **Assign Existing Material > road** from the marking menu; the *road* shader is applied to the plane.

6. Click on the *road* shader in the **Hypershade** window; the **road** tab is displayed in **Property Editor**. In this tab, choose the checker button corresponding to the **Color** attribute in the **Common Material Attributes** area of the **Property Editor**, refer to Figure 10-36; the **Create Render Node** window is displayed. Choose the **File** button from this window; the **File Attributes** area is displayed in the **file1** tab of the **Property Editor**.

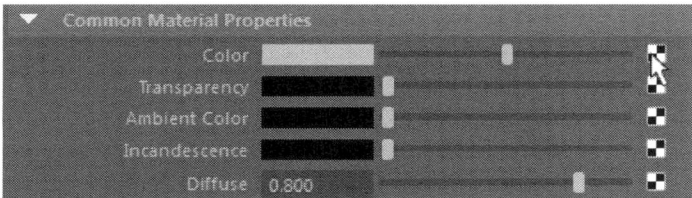

*Figure 10-36 The **Common Material Attributes** area*

7. Choose the folder icon available on the right of the **Image Name** attribute from the **File Attributes** area, refer to Figure 10-37; the **Open** dialog box is displayed. Next, select the **roadtexture.jpg** and then choose the **Open** button; the texture is applied to the *road*. Now, close the **Hypershade** window.

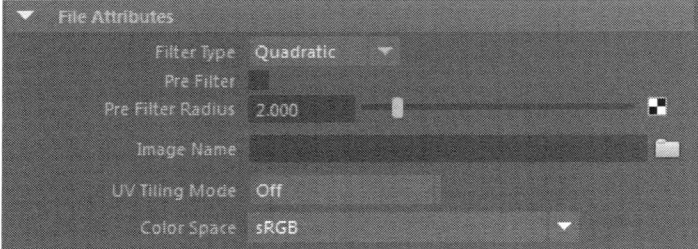

*Figure 10-37 The **File Attributes** area*

8. Switch to the top-Y viewport and press 6 to view the texture applied to the road. Make sure the plane is selected and then, choose **UV > Create > Planar > Option Box** to menubar

to open the **Planar Mapping Options** window. In this window, select the **Camera** option associated with the **Project from** attribute. Now, choose the **Project** button to enable planar projection on the plane.

9. Switch to persp viewport and then press 6 to view the texture applied to the road. In the **UV Coordinates** area of **Attribute Editor > file1** tab, choose the input button located at the right side of the **Uv Coord**; the properties of the coordinates are displayed.

10. In the **2d Texture Placement Attributes** area, make sure **1** is entered in the **Repeat UV** edit boxes. The plane after applying texture is shown in Figure 10-38.

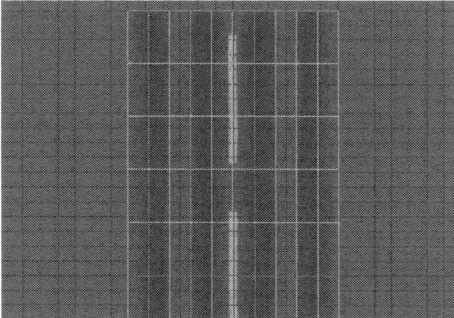

Figure 10-38 Texture applied on the plane

11. Choose **Create > Objects > Polygon Primitives > Cube** from the menubar. In **Channel Box / Layer Editor**, set the parameters of **polyCube1** in the **INPUTS** area, as shown in Figure 10-39. Next, duplicate **pCube1** and align both the cubes with the road to get a base for the street using the **Move Tool**, refer to Figure 10-40.

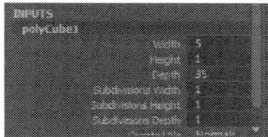
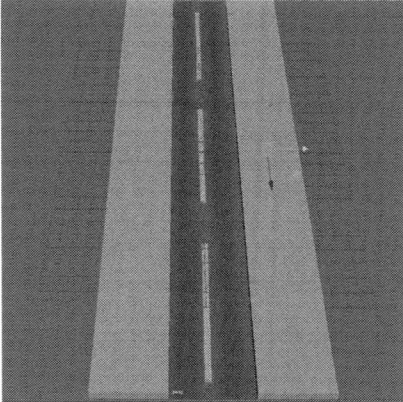

Figure 10-39 Setting the parameters of **polyCube1**

Figure 10-40 The base for the street displayed

Creating Buildings

In this section, you will create buildings by using the paint strokes.

1. Select the **Modeling** menuset from the **Menuset** dropdown list in the Status Line. Choose **Windows > Editors > General Editors > Content Browser** from the menubar; the **Content Browser** window is displayed. Choose the **Paint Effects** node, if it is not already chosen and then select the **cityMesh** folder in the left pane of the **Content Browser** window; the corresponding paint strokes are displayed in the right pane of the **Content Browser** window. Choose the **chicagoTower.mel** paint stroke from the **Content Browser** window, as shown in Figure 10-41. Next, close the **Content Browser** window.

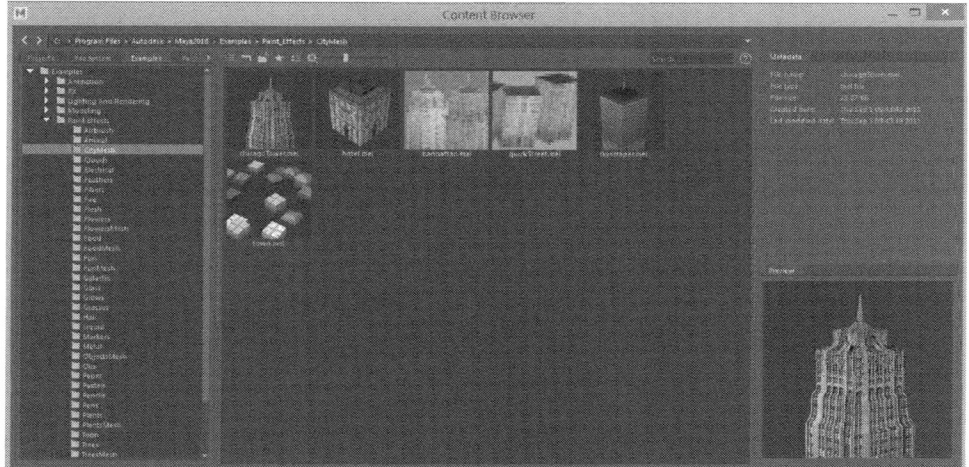

Figure 10-41 Choosing the chicagoTower.mel paint stroke from the Content Browser window

Note
You can adjust the attributes of the building paint stroke to get different results. Try using different cityMesh paint strokes from the Content Browser window to create different types of buildings.

2. Make sure the **Maya Classic** is selected in the **Workspace**. Next, choose **Generate > Template Brush Settings** from menubar; the **Paint Effects Brush Settings** window is displayed, as shown in Figure 10-42. Make sure the value **1** is set in the **Global Scale** edit box of this window to set the brush stroke. Close the **Paint Effects Brush Settings** window.

3. Maximize the top-Y viewport and press 6 to switch to the texture mode. Next, press the left mouse button and drag the cursor; buildings are displayed in the viewport, refer to Figure 10-43.

4. Make sure all building paint strokes are selected in the viewport. Choose **Edit > Duplicate > Duplicate Special > Option Box** from the menubar; the **Duplicate Special Options** window is displayed, as shown in Figure 10-44. In this window, enter **17** in the x edit box corresponding to the **Translate** attribute and then choose the **Duplicate Special** button; a duplicate of the building paint stroke is created and aligned to the opposite side of the plane. You might need to adjust the transformation value along the X-axis to align the building stroke.

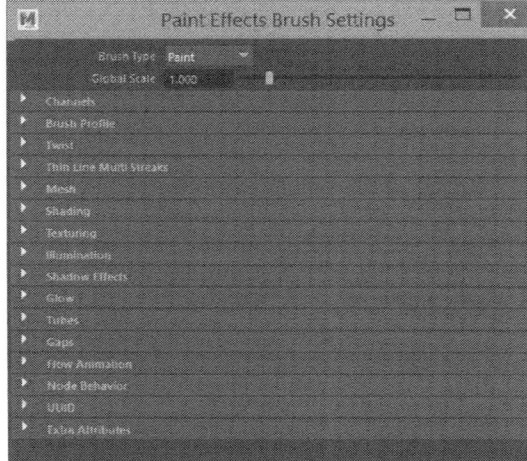

Figure 10-42 The **Paint Effects Brush Settings** window

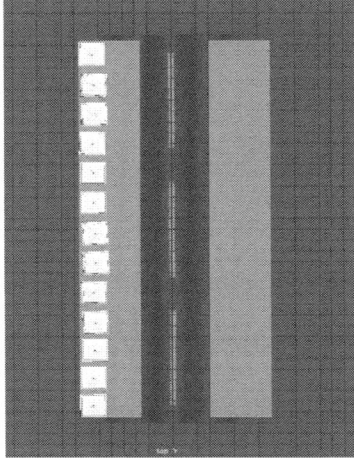

Figure 10-43 Buildings created

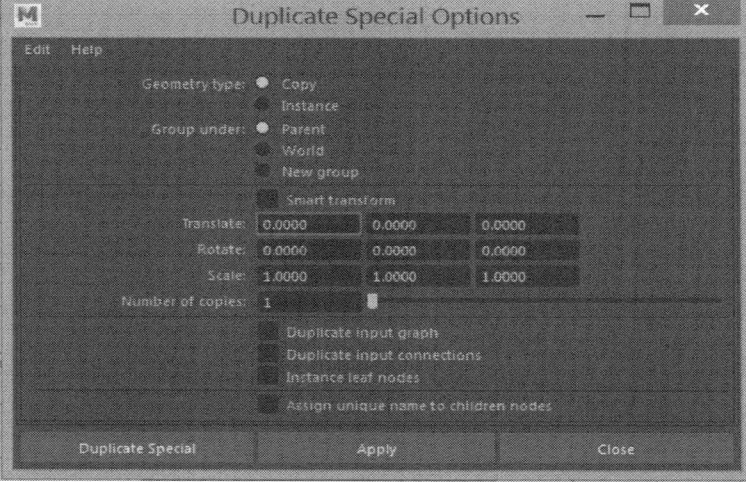

Figure 10-44 The **Duplicate Special Options** window

Paint Effects 10-19

5. Maximize the persp viewport. Figure 10-45 shows the building paint stroke created and aligned to the opposite side of the plane.

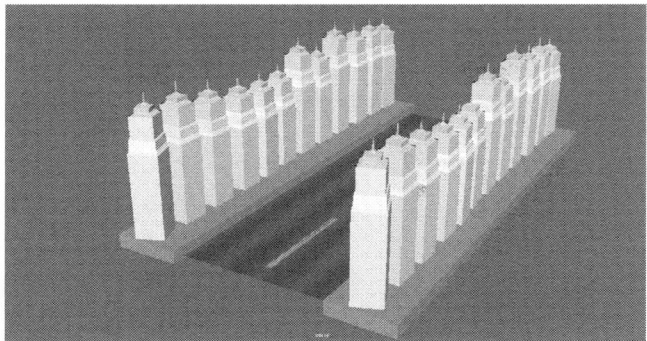

Figure 10-45 Building paint stroke created on the opposite side of the plane

Creating Clouds

In this section, you need to create clouds in the scene by using the paint strokes.

1. Choose **Windows > Editors > Outliner** from the menubar; the **Outliner** window is displayed. Next, select the **persp** camera in the **Outliner** window; various tabs in **Attribute Editor** are displayed.

2. Make sure the **perspShape** tab is chosen in the **Attribute Editor** and then expand the **Environment** area in it, refer to Figure 10-46. Next, click on the **Background Color** swatch in this tab; the **Color History** palette is displayed. Make sure the **HSV** option is selected in the drop-down list in the **Color History** palette. Now, enter the **HSV** values in the **Color History** palette, as shown in Figure 10-47.

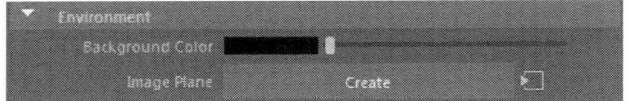

Figure 10-46 Attributes in the **Environment** area

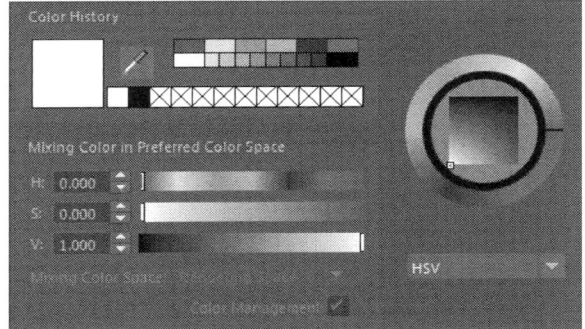

Figure 10-47 The **Color History** palette

3. Maximize the top-Y viewport and choose **Windows > Editors > General Editors > Content Browser** from the menubar; the **Content Browser** window is displayed. In the **Paint Effects** node of the **Content Browser** window, select the **cumulusPurple.mel** cloud type from the **Clouds** node. Next, in the top-Y viewport, press and hold the B key along with the left mouse button and then drag the cursor to the left or right to increase the brush size. Now, paint the cloud in the top-Y viewport, refer to Figure 10-48.

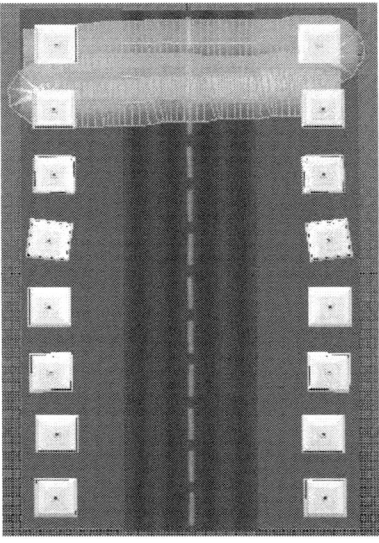

Figure 10-48 Cloud painted in the top-Y viewport

4. In the **Channel Box / Layer Editor**, enter **15** in the **Translate Y** and **-90** in the **Rotate X** edit boxes, respectively. Next, enter **100** in the **Global Scale** edit box in the **INPUTS** area of the **cumulusPurple1** in the **Channel Box / Layer Editor**.

5. Maximize the persp viewport and manually align the clouds paint stroke behind the buildings using **Move Tool**, refer to Figure 10-49.

6. Choose the **Render the current frame** button from the Status Line to render the scene in the Maya Software renderer.

 Note
You can create more instances of the clouds as per your requirement.

Paint Effects

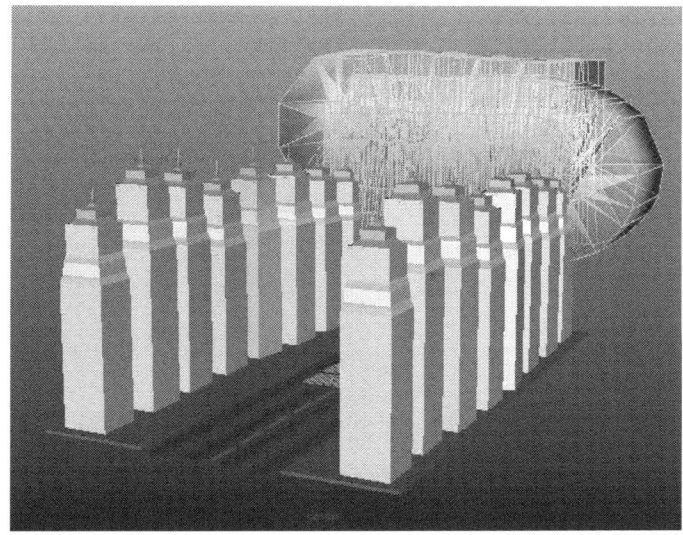

Figure 10-49 *Clouds aligned behind the buildings*

Creating Lights

In this section, you need to create lights to illuminate the scene.

1. Choose **Create > Objects > Lights > Ambient Light** from the menubar; the ambient light is created. Set the parameters of the light in the **Channel Box / Layer Editor**, as shown in Figure 10-50.

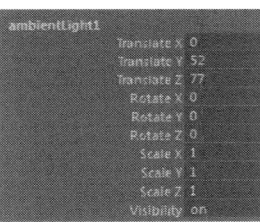

Figure 10-50 *The ambientLight1 parameters*

Saving and Rendering the Scene

In this section, you will save the scene that you have created and then render it. You can view the final rendered image of the scene by downloading the *c10_maya_2019_rndr.zip* file from *www.cadsofttech.com*. The path of the file is as follows: *Textbooks > Animation and Visual Effects > Maya > Autodesk Maya 2019 for Novices*.

1. Choose **File > Save Scene** from the menubar.

2. Choose **Create > Objects > Cameras > Camera and Aim** from the menubar; the camera is created in the viewport. Invoke the **Outliner** window and expand **camera1_group**. Next, select **camera1_aim** and enter values in the **Channel Box / Layer Editor** for setting the aim of the camera, as shown in Figure 10-51.

3. Select **camera1** from the **Outliner** window. Enter values in the **Channel Box / Layer Editor**

for setting the camera position, as shown in Figure 10-52. Next, close the **Outliner** window.

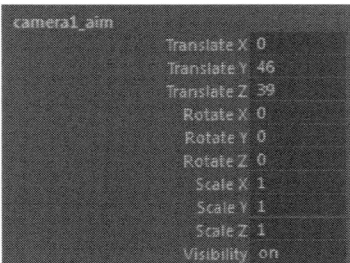 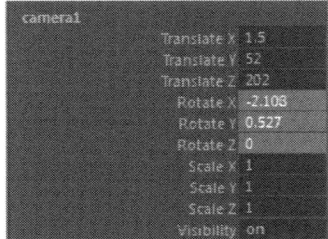

*Figure 10-51 The **camera1_aim** parameters*   *Figure 10-52 The **camera1** parameters*

4. In the **Attribute Editor**, make sure the **cameraShape1** tab is chosen. In the **Environment** area of this tab, choose the **Background Color** swatch; the **Color History** palette is displayed. In this palette, enter **1** in the **V** edit box.

5. Choose **Panels > Perspective > camera1** from the **Panel** menu; the scene view through the camera is displayed.

6. Choose the **Render the current frame** button from the Status Line to render the scene; the **Render View** window is displayed. This window shows the final output of the scene, refer to Figure 10-53.

Figure 10-53 The final output after rendering

EXERCISES

The rendered output of the models used in the following exercises can be accessed by downloading the *c10_maya_2019_exr.zip* file from *www.cadsofttech.com*. The path of the file is as follows: *Textbooks > Animation and Visual Effects > Maya > Autodesk Maya 2019 for Novices*.

Exercise 1

Extract the contents of *c10_maya_2019_exr.zip* and then open *c10_exr01_start.mb*. Now, use paint strokes to create an underwater scene around the ant model, as shown in Figure 10-54.

(Expected time: 30 min)

Figure 10-54 The underwater scene

Exercise 2

Create the model of a hut, as shown in Figure 10-55. Next, apply texture to it and create a tree on its left side by using the **Content Browser** window, refer to Figure 10-55.

(Expected time: 30 min)

Figure 10-55 The tree created on the left side of a hut model

Exercise 3

Create the model of a flower pot, as shown in Figure 10-56. Next, apply texture to it and use the **Content Browser** window to create flowers in the flower pot. Render the scene to get the final output, as shown in Figure 10-57. **(Expected time: 30 min)**

Figure 10-56 The flower pot   *Figure 10-57* The rendered flower pot

Chapter 11

Rendering

Learning Objectives

After completing this chapter, you will be able to:
- *Use the Render Setup*
- *Understand the basic concepts of rendering*
- *Use Arnold and Maya Software renderers*
- *Use Maya Hardware and Maya Vector renderers*

INTRODUCTION

Rendering is the process of generating a 2-dimensional image from a 3-dimensional scene. It is considered as the final stage in 3D production. Rendering helps in visualizing the lighting effects, materials applied, background, and other settings that you set for the scene. In Maya, you can create render layers and render the single layer or multiple layers using the **Render Layer Editor**.

RENDER SETUP

The **Render Setup** window is used to create, edit, and delete layers. It is also used to control the layer bends, collections, and overrides. To open the **Render Setup** window, refer to Figure 11-1, choose the **Launch Render Setup** button from the Status Line; the **Render Setup** window will be displayed. The **Render Setup** window is divided into two tabs: **Render Setup** and **Property Editor-Render Setup**. The **Render Setup** tab allows you to create layers, collections, and overrides whereas the **Property Editor-Render Setup** tab allows you to set corresponding values.

> **Tip**
> *If you use the **Rendering - Standard** or **Rendering - Expert** workspaces, the **Render Setup** and **Property Editor-Render Setup** tabs are automatically docked for you.*

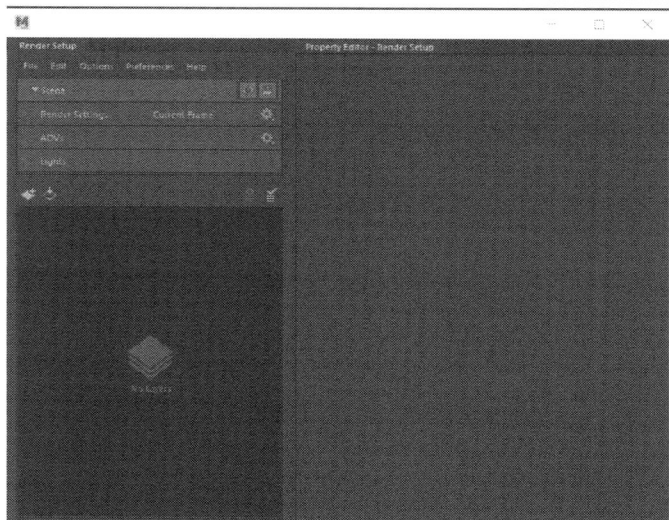

*Figure 11-1 The **Render Setup** window*

In Maya, different types of renderers are used to get the final output of a scene. Some of the most common renderers are discussed next.

MAYA SOFTWARE RENDERER

The **Maya Software** renderer is the default renderer in Maya. It is an advanced, multi-threaded renderer that produces high quality images. This renderer is used to produce effects such as advanced shadows and reflections. It also supports most of the entities in Maya such as particles, fluid effects, and paint effects.

Rendering

The **Maya Software** renderer has an advanced feature called **IPR** which stands for Interactive Photorealistic Rendering. It is used to preview and make interactive adjustments in the rendered image. It creates a special image file that not only stores the pixel information of an image, but also the data of the surface normals, materials, and objects associated with each of these pixels. Maya updates this information in the **Render View** window as you make changes to the shades or lighting of the scene.

MAYA HARDWARE RENDERER

The **Maya Hardware** renderer is an efficient renderer and it can render depth map shadows. It uses graphics buffers and memory of the computer to generate renders. It also has limitations as it does not render ray trace shadows, reflections, or post-process effects like glow. Particles are rendered using the **Maya Hardware 2.0** renderer for the alpha information. If you are rendering a scene for the first time using the **Maya Hardware 2.0** renderer, it may take more time. It is so because, the scene is first converted into a data structure, and then it will be calculated by the graphic division of the CPU. The **Maya Hardware 2.0** renderer uses the same tessellation settings that are used in the **Maya Software** renderer. Various settings of the **Maya Hardware 2.0** renderer are discussed next.

The Maya Hardware Renderer Settings

Choose **Windows > Editors > Rendering Editors > Render Settings** from the menubar; the **Render Settings** window will be displayed. Select **Maya Hardware 2.0** from the **Render Using** drop-down list. Next, choose the **Maya Hardware 2.0** tab from the **Render Settings** window; the hardware render settings will be displayed, as shown in Figure 11-2.

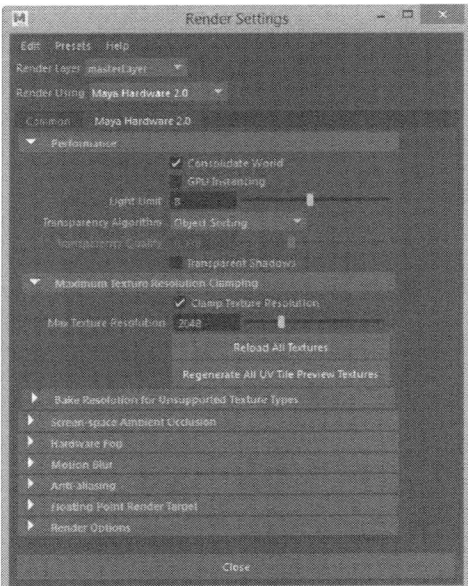

Figure 11-2 The partial view of hardware renderer settings displayed

MAYA VECTOR RENDERER

The **Maya Vector** renderer is used to create unrealistic images such as cartoons, tonal art, wireframe, motion blur, and so on, shown in Figure 11-3. Such rendered images can be saved in various formats such as *swf, ai, tiff, svg,* and *eps.*

One of the vector rendered images is shown in Figure 11-4. In this figure, you will notice that the image is more stylized and has more color effects as compared to the images created by using other renderers.

 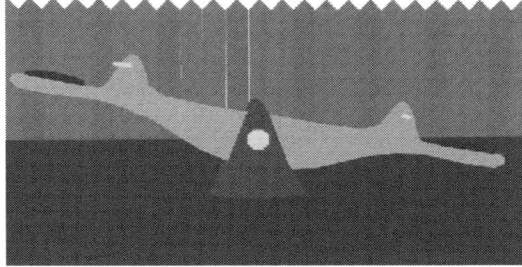

Figure 11-3 An object with the motion blur effect   *Figure 11-4* The vector rendered image

To understand the process of creating effects using the **Maya Vector** renderer better, compare Figure 11-5 with Figure 11-6. In these figures, you will notice that Figure 11-5 is more realistic than the one shown in Figure 11-6. This is because Figure 11-6 displays tonal art effect and such effect is mainly used for creating logos and diagrams.

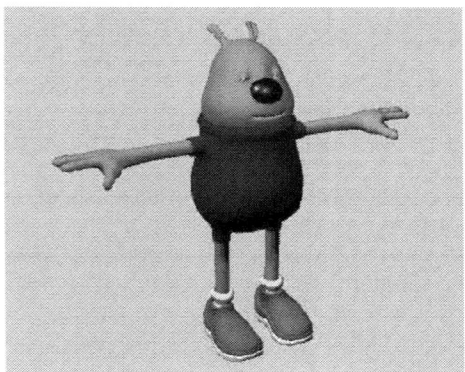 

Figure 11-5 An image rendered using the **Maya Software** renderer   *Figure 11-6* An image rendered using the **Maya Vector** renderer

The **Maya Vector** renderer is based on the concept of the RAViX technology. RAViX stands for Rapid Visibility Extension. This technology converts a 3D model into a 2D vector-based image by detecting the lines and vertices that make up a 3D model and then converts them into shaded polygons for recreating the 3D image in a 2D vector format (specifically Adobe Flash SWF and EPS formats). RAViX provides per polygon shading capabilities that are superior to other rendering technologies. Also, the file size created by using this technology is smaller than the other technologies. The **Maya Vector** renderer does not support any light, except the point light. If there is no light in a scene, the vector renderer creates a default light at the camera position during rendering.

Rendering

The **Maya Vector** renderer cannot render some features such as bump maps, displacement maps, Maya fluid effects, image planes, Maya fur, multiple UVs, Maya Paint Effects, particles, post-render effects, and textures. To render an object with any of these features using the **Maya Vector** renderer, first you need to convert the object into a polygon and then render the object. To convert an object into a polygon, select the object from the viewport and then choose **Modify > Objects > Convert** from the menubar; a cascading menu comprising of all tessellation methods will be displayed. Choose the required option from the cascading menu; the object will be converted into a polygon.

The Maya Vector Renderer Settings

Choose **Windows > Editors > Rendering Editors > Render Settings** from the menubar; the **Render Settings** window will be displayed. In this window, select **Maya Vector** from the **Render Using** drop-down list. Next, choose the **Maya Vector** tab from the **Render Settings** window; the vector renderer settings will be displayed, as shown in Figure 11-7.

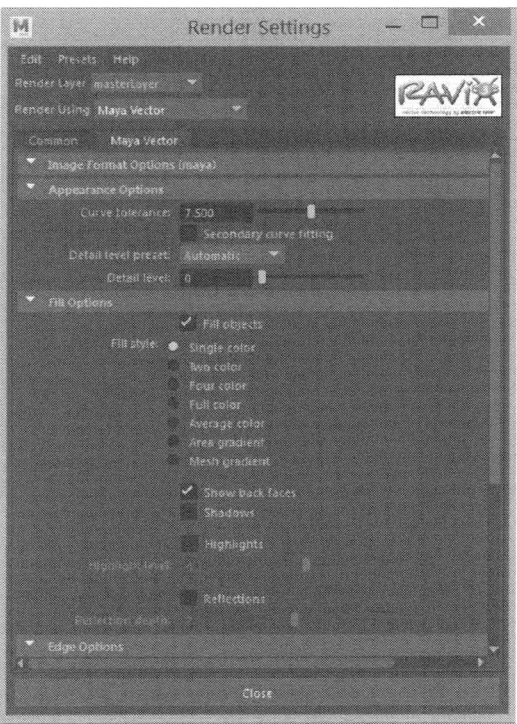

Figure 11-7 The partial view of vector renderer settings displayed

If the **Maya Vector** option is not available in the **Render Using** drop-down list, choose **Windows > Editors > Setting/Preferences > Plug-in Manager** from the menubar; the **Plug-in Manager** window will be displayed. Select the **VectorRender.mll** check box, refer to Figure 11-8. Next, choose the **Refresh** button and then the **Close** button; the **Maya Vector** option will be displayed in the **Render Using** drop-down list.

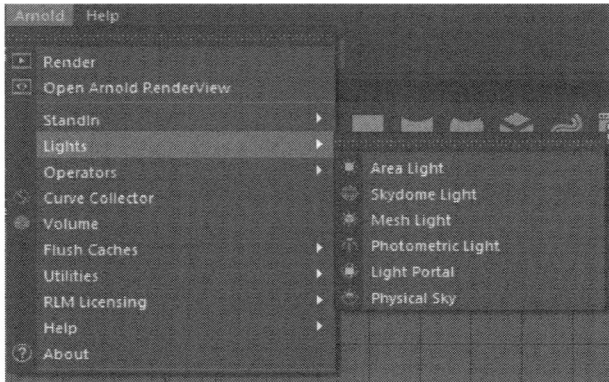

*Figure 11-8 The **Arnold** menu*

ARNOLD RENDERER

Arnold renderer is a Maya plug-in and also known as MtoA. Arnold plug-in allows you to use the Arnold Renderer directly in Maya. Arnold is a cross-platform rendering solution developed by Solid Angle. It is used by prominent studios in the animation, broadcast, and gaming industries across the globe.

The Arnold renderer takes a different approach than the renderers that use biased algorithms such as photon mapping or final gather. Such algorithms cache the data and then re-sample it later. In the process, they take large amount of memory and introduce artifacts such as sampling artifacts. Arnold is an unbiased rendering engine and uses a physically-based Monte Carlo ray/path tracing algorithm. It does not use any caching algorithm thus produces clear and photo-realistic renders.

By default, the Arnold Renderer is active in Maya. If it is not loaded automatically, choose **Windows > Editors > Setting/Preferences > Plug-in Manger**; the **Plug-in Manager** window will be displayed. In this window, select the **Loaded** and **Auto Load** check boxes corresponding to the **mtoa.mll** entry and then choose the **Close** button.

WORKING WITH LIGHTS

You can use regular Maya lights with Arnold. However, the **Ambient** and **Volume** lights are not supported by Arnold. You can create Maya lights from the **Create** menu as well as from the **Arnold** menu, refer to Figure 11-8. Arnold also has its own custom lights. Both type of light are discussed next.

Working with Maya Lights

Setup a scene and then choose **Create > Objects > Lights > Point Light** from the menubar; a point light will be created in the viewport. Adjust the position of the light, refer to Figure 11-9. If you render the scene, you will see darker output because of the fall of type of the light which is set to quadratic by default. You will know about the decay type later in this section.

Rendering

*Figure 11-9 The position of the **Point Light** in the scene*

You can use Arnold RenderView to view the render of the scene. To open it, choose **Render** from the **Arnold** menu. When you create a light in Maya, the **Arnold** area appears in the **Attribute Editor**, using the options from this area, you can change Arnold specific attributes of the light. Some light attributes are discussed next.

Arnold has some built-in custom lights such as **Area Light**, **Skydome Light**, **Mesh Light**, **Photometric Light**, **Light Portal**, and **Physical sky**. You can create these lights from the **Arnold** menu. Commonly used lights are discussed next.

Area Light

When you use the regular Maya area light, Arnold considers it a rectangular or quad light source but when you use the Arnold's **Area Light**, you have options for specifying another shape for the light source from the **Light Shape** drop-down list, refer to Figure 11-10. Also, note that the **Intensity** and **Exposure** attributes appear together in the **Attribute Editor**.

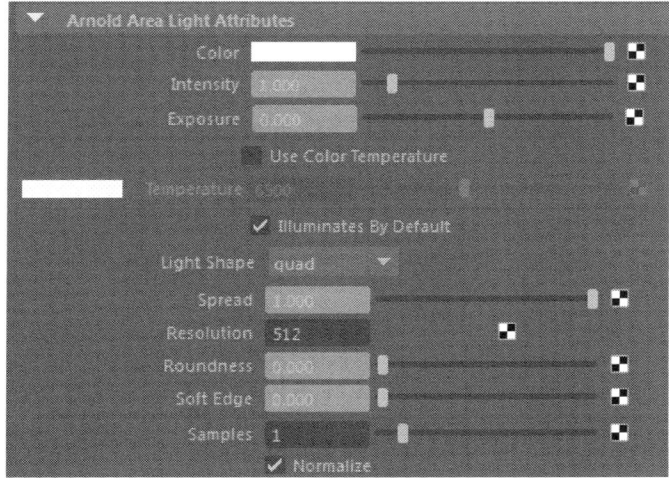

*Figure 11-10 The **Light Shape** drop-down list*

When a shader is connected to an Area Light, Arnold calculates the importance tables for efficient sampling according to the luminance values of the textures. The resolution of the tables is controlled by the **Resolution** attribute. For best result, you should match attribute's value with the resolution of the incoming image.

The **Spread** attribute controls the focus of the light in the direction along the normal. At a value of 1, light will be focussed like a laser beam.

Mesh Light
You can use the **Mesh Light** to create light shapes which are not possible to create from regular shapes such as cylinder or rectangle. To create a mesh light, select a mesh in the viewport and then choose **Arnold > Lights > Mesh Light** from the menubar. To set the attributes of the light, choose the shape node of the mesh in the **Attribute Editor** and then expand the **Arnold** area. Make sure that the **mesh_light** option is selected from the **Arnold Translator** drop-down list.

Photometric Light
This light is used to import and render real-world light distribution files, IES files. An IES file contains the measurement of the light intensity stored in an ASCII file. When you create a photometric light, you can import the file by clicking the folder icon corresponding to the **Photometry File** attribute from the **Photometric Light Attributes** area of the **Attribute Editor**.

STANDARD SHADER
To create any type of material from wood to plastic, from chrome to aluminium, and so on, you can use Arnold's Standard Shader. It is also known as Ai Standard Surface. Once you apply the shader to a mesh, you can control attributes from the **Attribute Editor**. This shader has large number of controls which are grouped under different areas in the **Attribute Editor**. The commonly used areas are discussed next.

Base
The **Color** attribute is used to set the brightness of the surface when lit directly by a white light source. It defines the percentage of each component of the spectrum which is not absorbed by the surface. The **Weight** attribute defines the weight of the diffuse component. The **Roughness** attribute controls the roughness of the surface. Higher values are suitable for creating material like plaster or sand. The **Metalness** attribute controls the metallic effect to create the metallic surface. Higher values are suitable for creating material like metal.

Specular
The attributes in this area are used to control the direct and indirect reflections. You can also make the reflections blurry. The **Color** control defines the color of the reflection. You can use this attribute to tint the reflections. The **Weight** attribute controls the brightness of the specular highlights. The **Roughness** attribute controls the glossiness of the specular highlights. The **Anisotropy** attribute reflects and transmits light in a direction that causes the surface to look shiny or rough in a certain direction. The **Rotation** attribute controls the orientation of the anisotropic highlights. The IOR attribute defines the index of refraction of the medium.

Rendering

Transmission

The **Weight** parameter controls the amount of light that will scatter through the surface. This is useful in creating materials like glass and water. The **Color** attribute defines the color of the glass. The longer light penetrates the mesh, darker the color will be. The **Depth** attribute controls the depth into the volume at the transmission color takes place. The **Scatter** attribute defines the scattering effect. The **Scatter Anisotropy** attribute controls the anisotropy of the scattering. The **Dispersion Abbe** attribute specifies the Abbe number of the material. This attribute is suitable for creating surfaces like diamond. The **Extra Roughness** attribute adds additional blurriness to the surface.

Bump Mapping

The **Bump Mapping** attribute in the **Geometry** area allows you to connect a shader to it. The shader affects the normals of the surface to create the bump effect.

Emission

The attributes in this area allow you to create self-illuminating surfaces. The **Color** attribute defines the emitted light color. The **Weight** attribute defines the amount of emitted light on the surface.

Matte

The attributes in this group allow you to create cutouts by rendering the alpha as 0. Select the **Enable Matte** check box to enable the matte effect. The **Matte Color** attribute defines the color of the matte and the **Matte Opacity** attribute defines the opacity of the cutout.

TUTORIAL

All the files used in the tutorials can be downloaded from the CADSoft website (*www.cadsofttech.com*). These files are compressed in zip file format and are required to be extracted before using them in the tutorials. The path of the files is as follows: *Textbooks > Animation and Visual Effects > Maya > Autodesk Maya 2019 for Novices*

Tutorial 1

In this tutorial, you will create a simple studio setup using Arnold lights, as shown in Figure 11-11. **(Expected time: 30 min)**

The following steps are required to complete this tutorial:

a. Create a project folder.
b. Download and open the file.
c. Add lights to the scene.
d. Create material.
e. Save the scene.

*Figure 11-11 The rendered image of a geometry using the **Arnold** renderer*

Creating a Project Folder
Create a new project folder with the name *c11_tut1* at *\Documents\maya2019*, as discussed in Tutorial 1 of Chapter 2.

Downloading and Opening the File
In this section, you need to download and open the file.

1. Download the *c11_maya_2019_tut.zip* from *www.cadsofttech.com*. The path of the file is as follows: *Textbooks > Animation and Visual Effects > Maya > Autodesk Maya 2019 for Novices*.

   Next, extract the contents of the zip file to the *Documents* folder.

2. Choose **File > Open Scene** from the menubar; the **Open** dialog box is displayed. In this dialog box, browse to *\Documents\c11_maya_2019_tut* and select **c11_tut1_start.mb** file in it. Choose the **Open** button; the file opens.

3. Now, choose **File > Save Scene As** from the menubar; the **Save As** dialog box is displayed. As the project folder is already set, the path *\Documents\maya2019\c11_tut1\scenes* is already displayed in the **Look In** drop-down list. Save the file with the name *c11tut2.mb*.

Adding Lights to the Scene
In this section, you will add Arnold lights to the scene.

1. Choose **Arnold > Lights > Area Light** from the menubar; an area light is added to the scene. Place the light at the left of the geometry, as shown in Figure 11-12. Render the scene; you will notice that the render is dark.

   Next, you will adjust the light's properties.

Rendering

Figure 11-12 *The Arnold light placed in the scene*

2. In the **Attribute Editor > aiAreaLightShape1 > Arnold Area Light Attributes** area, enter **4** and **8** in the **Intensity** and **Exposure** edit boxes, respectively. Render the scene. Now the render looks brighter, as shown in Figure 11-13.

Figure 11-13 *The render of the scene*

3. Duplicate the light and place it on the right side of the geometry. Again, duplicate the light and place it at the top of the geometry, refer to Figure 11-14.

Figure 11-14 *The other Arnold light placed in the scene*

4. In the **Attribute Editor > aiAreaLightShape1 > Arnold Area Light Attributes** area, enter **3** in the **Samples** edit box. Repeat the process for other two lights.

5. In the **Attribute Editor > aiAreaLightShape1 > Arnold Area Light Attributes** area, select the **Use Color Temperature** check box.

6. In the **Attribute Editor > aiAreaLightShape2 > Arnold Area Light Attributes** area, select the **Use Color Temperature** check box and then enter **4000** in the **Temperature** edit box. Render the scene; notice that the warm and cool temperatures are producing a nice studio light setup.

Creating the Material

In this section, you will create a material for the scene using the **Standard** shader.

1. Right-click on the geometry in the scene and then choose **Assign New Material** from the shortcut menu; the **Assign New Material** window is displayed. In this window, choose **Arnold > Shader > Surface > aiStandardSurface**; the Standard shader is applied to the geometry.

2. In the **Attribute Editor > aiStandardSurface1** tab, choose the **Presets** button; a flyout is displayed. Choose **Brushed_Metal > Replace** from the flyout to apply preset to the geometry. Render the scene; you will notice that there is some noise in the specular highlights, refer to Figure 11-15. The noise occurs because of low samples.

Figure 11-15 The render with noise in the specular highlights

Next, you will adjust samples.

4. Choose **Display render settings** button on the Status Line; the **Render Settings** window is displayed. In the **Sampling** area of the **Arnold Renderer** tab, enter **3** and **4** in the **Diffuse** and **Specular** edit boxes, respectively. Render the scene.

Saving the Scene

In this section, you need to save the scene that you have created.

Rendering

1. Choose **File > Save Scene** from the menubar to save the scene.

EXERCISES

The rendered output of the scenes used in the following exercises can be accessed by downloading the *c11_maya_2019_exr.zip* file from *www.cadsofttech.com*.. The path of the file is as follows: *Textbooks > Animation and Visual Effects > Maya > Autodesk Maya 2019 for Novices.*

Exercise 1

Create a scene, as shown in Figure 11-16. Apply textures to the scene and then render it using the **Arnold** renderer to get the output shown in Figure 11-17. **(Expected time: 45 min)**

Figure 11-16 Scene before rendering

Figure 11-17 Scene after rendering

Exercise 2

Extract the contents of the *c11_maya_2019_exr.zip file*. Open *c11_exr02_start.mb* and then apply textures to it. Next, create a tree on its left using the **Content Browser** window, as shown in Figure 11-18. Next, add lights and render the scene using the **Arnold** renderer to get the output shown in Figure 11-19. **(Expected time: 30 min)**

Figure 11-18 The tree created in the scene

Figure 11-19 The rendered scene

Index

A

Aim constraint 9-13
Animation Layer Pane 8-15
Animation preferences button 1-14, 8-3
Arnold Renderer 11-6
Attach tool 4-4
Auto keyframe toggle 1-14

B

Ball Joint 9-3
Bend deformer 9-7
Bifrost Aero material 6-6
Bifrost Foam material 6-7
Blend Shape deformer 9-5
Bones and Joints 9-2
Boundary tool 3-12
Brush Profile area 10-6
Bump Mapping attribute 11-9

C

Camera-Based mapping 5-5
Center Pivot option 1-9
Color attribute 11-8
Content Browser 10-2
Contour Stretch mapping 5-5
Curve degree 3-7
Curve Warp deformer 9-5
Cylindrical mapping 5-3

D

Duplicate NURBS Patches 4-2

E

Emission area 11-9
Enable normalized curve display 8-12
Enable stacked curve display 8-12, 8-13
Extrude tool 3-11

F

Flatness 1 and Flatness 2 attributes 10-6
Frame Rate 8-4
FX menuset 1-5

G

Global Scale attribute 10-5
Graph Editor 8-8

H

Hardware Renderer 11-3
Hinge Joint 9-4
Hypershade window 6-2

I

Illumination area 10-9
Insert Keys Tool 8-8

K

Keyframe Animation 8-2
Kinematics 9-4

L

Linear tangents tool 8-10
Loft tool 3-9

M

Matte area 11-9
Maya Software Renderer 11-2
Maya Vector Renderer 11-4
Menuset drop-down list 1-6

N

Nonlinear Animation 8-2
Normal-Based mapping 5-5

O

Open/Close tool 4-6
Open the Dope Sheet tool 8-14

P

Panel menu 1-15
Parent and Child relationship 9-4
Path Animation 8-4
Planar mapping 5-5
Playback Controls 8-3
Post-Infinity Cycle tool 8-13
Post-infinity cycle with offset tool 8-13
Pre-infinity cycle with offset tool 8-13
Project Curve on Surface tool 4-2
Property Editor 6-4
Push option 4-7

R

Ramp Shader 6-9
Reflection area 11-9
Renderer 1-16
Render Setup window 11-2
Right-click Menus viii

S

Select Tool 1-8
Shaded tool 5-3
Shading area 10-9

Shadow Diffusion attribute 10-11
Shadow Offset attribute 10-11
Shadow Transp attribute 10-12
Sine deformer 9-9
Softness attribute 10-6
Specular area 11-8
Spherical mapping 5-4
Square tool 3-12
Squash deformer 9-9
Standard Shader 11-8

T

Technical Animation 8-3
Tool Box 1-8
Transmission area 11-9
Twist area 10-7
Twist Rate 10-7
Types of Joints 9-3

U

Universal Joint 9-4
UV Distortion tool 5-3
UV Editor 5-2
UV mapping 5-2
UV Toolkit tool 5-3

V

Value snap on/off tool 8-12
View Toolbar 5-3

W

Wireframe tool 5-3
Workspaces 1-22
Wrap deformer 9-7

Manufactured by Amazon.ca
Bolton, ON